IntechOpen

Advances in Microfluidics and Nanofluids

Edited by S. M. Sohel Murshed

Published in London, United Kingdom

Supporting open minds since 2005

Advances in Microfluidics and Nanofluids
http://dx.doi.org/10.5772/intechopen.91560
Edited by S. M. Sohel Murshed

Contributors
Susanne Seibt, Timothy Ryan, Ronald Nguele, Katia Nchimi Nono, Kyuro Sasaki, Mehdi Mohammadi, Yousef Alihosseini, Amir Rezazad Bari, Nan Zhang, Honggang Zhang, Haoyang Zhang, Tianyu Guan, Xiangyu Wang, Johann F. Osma, Olga Patricia Fuentes Daza, Juliana Noguera, Paula Peñaranda, Sergio Florez, Juan Carlos Cruz, Liji Huang, Abang Annuar Ehsan, Muhammad Syafiq Rahim, Meijie Chen, Xingyu Chen, Dongling Wu, S. M. Sohel Murshed

First published in London, United Kingdom, 2021 by IntechOpen
IntechOpen is the global imprint of INTECHOPEN LIMITED, registered in England and Wales, registration number: 11086078, 5 Princes Gate Court, London, SW7 2QJ, United Kingdom
Printed in Croatia

British Library Cataloguing-in-Publication Data
A catalogue record for this book is available from the British Library

Additional hard and PDF copies can be obtained from orders@intechopen.com

Advances in Microfluidics and Nanofluids
Edited by S. M. Sohel Murshed
p. cm.
Print ISBN 978-1-83968-692-4
Online ISBN 978-1-83968-693-1
eBook (PDF) ISBN 978-1-83968-694-8

For EU product safety concerns:
IN TECH d.o.o., Prolaz Marije Krucifikse Kozulić 3, 51000 Rijeka, Croatia, info@intechopen.com or visit our website at intechopen.com.

Meet the editor

Prof. S. M. Sohel Murshed was born in Bangladesh and obtained his Ph.D. in Mechanical and Aerospace Engineering from Nanyang Technological University, Singapore. He is currently a professor at the Mechanical Engineering Department of the University of Lisbon, Portugal. Previously, he worked as a postdoctoral fellow and visiting professor and scientist at different universities in Singapore, the USA, the UK, and India. He received the prestigious DUO-India Professorial Fellowship Award in 2020. Prof. Murshed has authored/co-authored 8 books, 28 book chapters, and over 160 papers in leading international journals and conferences. A number of his papers have been classified as Highly Cited Papers and his current citation is 6850 with an h-index of 30. Prof. Murshed was recently named in the list of the World's Top 2% Scientists.

Contents

Preface

Both microfluidics and nanofluids are rapidly growing technologies that have emerged as multidisciplinary research fields. The impact and importance of these two technologies can be manifested from their tremendous potentials and diverse applications such as virus detection, cell manipulation and separation, 3D printing, anticancer drug screening, advanced thermal management, and energy harvesting and storage. This book covers eclectic areas of these emerging technologies starting from their fundamentals to development to applications and it is composed of nine chapters that are organized into two sections: one related to microfluidics and the other on nanofluids and cooling.

The first chapter presents an overview of advances in both microfluidic and nanofluid technologies. This chapter provides the overall information and discusses the progress of these technologies. In addition to reviewing their key features, their applications and challenges are also highlighted.

In the second chapter, we discuss the applications of microfluidic devices specifically developed for the investigation of time-resolved analysis of growth kinetics and the structural evolution of nanoparticles and nanofibers. Focus is placed on the design considerations required for spectrometry and SAXS analysis. This chapter also discusses the use of these devices for time-resolved research.

The third chapter demonstrates commonly used manufacturing technologies and the process chain for prototyping and mass production of microfluidic chips. It details various types of rapid prototyping technologies besides presenting some important research findings. It also provides good guidance from processing materials and method selection for chip production to end-users.

A review of microfluidic flow sensing approaches is presented in Chapter 4. It covers numerous aspects including currently available products, microfluidic flow sensing technologies, major factors impacting flow metrology, and prospective sensing approaches for future microfluidic flow sensing.

Chapter 5 provides a comprehensive review of the fabrication of circular and rectangular cross-section channels of microfluidic devices using micromilling process. Various process and machining parameters are also discussed in this chapter.

The application of micromixers for wastewater treatment and an assessment of their life cycle are presented in Chapter 6. Six different micromixer designs and an evaluation of the performance of each in the treatment of wastewater are discussed. Their performance in terms of environmental impact was assessed through the life cycle assessment (LCA) methodology.

In the second section of this book, three contributions to nanofluids and cooling are included. The first one (Chapter 7) is on the fundamentals and applications of solar thermal conversion of plasmonic nanofluids. It summarizes the preparation methods of plasmonic nanofluids and reviews the solar absorption performance of these nanofluids based on the theoretical and experimental design.

The recent advancements of nanocomposite and nanofluids toward sustainable carbon capture, utilization, and storage are presented in Chapter 8. This chapter focuses primarily on nanomaterial applications for both fossil and renewable energies.

The last chapter deals with the effective parameters for increasing the efficiency of heat sinks and the application of liquid cooling in real life. Recent advances in developing an efficient heat sink including different parameters like geometry and flow parameters are reviewed. It also highlights the current gap between academic research and industry.

The book is expected to be a source of information for different communities including students, researchers, manufacturers, academicians, and professionals of these fields.

I would like to thank all the authors for their high-quality contributions and the publishing team for their cooperation and support.

I dedicate this book to my parents and siblings for their sacrifices and support. I owe my success to them.

S. M. Sohel Murshed
Instituto Superior Técnico,
University of Lisbon,
Lisbon, Portugal

Chapter 1

Introductory Chapter: An Overview of Advances in Microfluidics and Nanofluids Technologies

S.M. Sohel Murshed

1. Introduction

Both microfluidic and nanofluids are rapidly growing and hugely potential technologies which emerged as multidisciplinary research fields. While microfluidic technology began with the development of the first lab-on-a chip in 1979 microfluidic research had its first step forward only when Manz introduced the idea of μ-TAS (Micro total analysis systems) in 1990 [1]. Other main timelines of the development of microfluidic technology include introduction of microfluidics in cell biology and biochemistry in 1994, employing PDMS (polydimethylsiloxane) in microchips production in 1998, introduction of digital microfluidics in 2000, investigations on microfluidics cell culture systems in 2004, development of organ-on-chip technology between 2005 and 2010, emergence of paper-based microfluidics in 2007, application of 3D printing in microfluidic technology in 2010 and the latest development of microfluidics for theranostics between 2012 and 2015 [2, 3]. Microfluidic technology deals with small amounts (i.e., microliter or nanoliter) of fluids (liquids and gases) in micron or sub-micron size devices/systems/geometries. Microfluidics is a very popular research field which can be evidenced from the rapid growth of numbers of publications on this topic (**Figure 1**). Year-wise publication data on microfluidics for the past two decades (data obtained by searching topic: "microfluidic" in Web of Sciences on 10th May 2021) are presented in **Figure 1**. The impact and importance of microfluidic technology can be manifested from its diverse real-world applications ranging from virus detection and bioanalytical, cell manipulation and separation, 3D printing, paper microfluidics to anticancer drug screening as highlighted in the following section. The advantages of microfluidics include very small quantity (e.g., microliter) of sample or reagent usage, contamination risk reduction, low cost (e.g., for analysis/diagnosis), automation, enhanced sensitivity, accuracy and reliability. Although different materials have been used, PDMS is currently the most widely used material for the fabrication of microfluidic devices and systems.

On the other hand, nanofluids, which is a new class of heat transfer fluids coined in early 1990's [4, 5] have also attracted tremendous attention from the researchers due to their enhanced thermophysical properties, potential benefits, and numerous applications [6–9]. **Figure 1** also shows publication records on nanofluids during the last two decades and the number of publications on nanofluids (search by the topic: Nanofluids) increases exponentialy. It is to note that nanofluid is a much smaller

IntechOpen

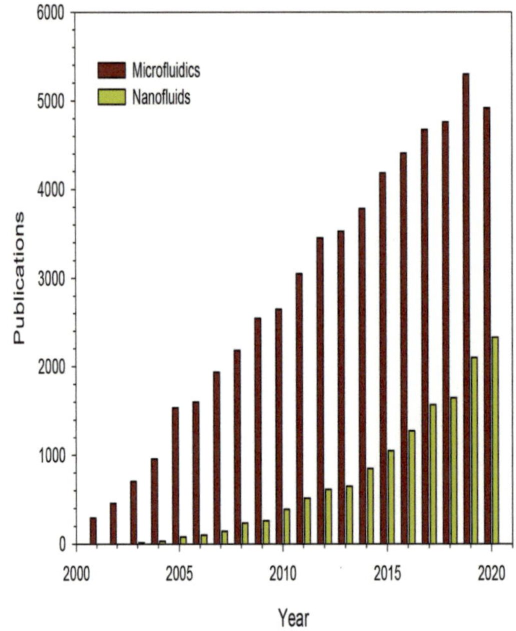

Figure 1.
Microfluidics and nanofluids related publications from 2001 to 2020 (source: Web of science).

field of research without any (noticeable) real-world applications compared to microfluidics. Nonetheless, these numbers indicate research activities and popularity of both research fields. Another important aspect of nanofluids is that having superior properties and nano-sized particles, they can be applied to microfluidic systems and devices which can result in improving performance and diversifying the applications of both nanofluids and microfluidic technologies [10, 11]. Nanofluids also span a wide range of potential applications starting from thermal management, energy conversion to nuclear reactor. However, this field is far from developing, exploring its benefits, and exploiting its real applications. The main challenges of nanofluids research are their inconsistent data, unidentified underlying mechanisms for the observed results and maintaining their long-term stability. The progress of the nanofluids research is briefly overviewed in the following section.

Since both microfluidics and nanofluids fields are well established and reported in the literature and textbooks [2, 3, 6–9, 12, 13], they will not be elaborated further. However, advances, applications and challenges of these technologies are highlighted in this chapter.

2. Advances in microfluidics

2.1 Global market status

Microfluidic technology is not only attractive to researchers and academics as an emerging research field but also to industrial people as its market is growing rapidly. According to a report by Yole development [14] currently there are 700+ microfluidic related companies worldwide yielding product revenues of around 7 B (billion) dollars in 2017. Also, there are 4500+ published patents only on microfluidic technology-based diagnostics. It was also forecasted microfluidic product market value to reach about 25 B dollars in 2025 [14]. The major contribution to this revenue is expected to come from new areas like clinical and point-of care diagnostics through microfluidic chips. Other areas such as optical actuations and drug delivery are also growing and are anticipated to contribute considerably to this huge market.

Recent research and development at the cutting edge of microfluidics technology span from 3D printing to virus detection. It is noted that microfluidic technology received more attention during COVID-19 pandemic due their potential in detection and diagnosis of this novel SARS-CoV2 virus. Common classifications of microfluidic technology are presented in **Figure 2**. Although continuous flow based microfluidic systems particularly for reaction and mixing are widely used, during the past two decades significant progress has also been made in research and development in droplet-based microfluidics [15]. Each of these main and subclassifications of microfluidics technology is well-developed or in good progress towards revolutionizing their respective area.

2.2 Development in application

During the past three decades applications of microfluidic technology increased considerably in a broad spectrum of scientific areas from biomedical (known as biomicrofluidics), chemical to MEMS thermal management. The most notable applications include immunoassays (a bioanalytical technique) used in pharmaceutical and clinical laboratories for diagnostics, DNA assays (capturing separation and detection of DNA etc), cell-based assays (known cell culture). A schematic of presentation of various fields of applications of microfluidic technology is provided in **Figure 3**. It can be seen that microfluidics really span a diverse field of applications starting from Lab-on-chip (LOC) to the food and agriculture sectors.

As a pioneering effort by the author nanofluids were studied in microfluidic geometries particularly in droplet-based microfluidics in order to explore new applications of nanofluids [10, 11, 16]. Thus, experimental investigations on the droplet formation and size manipulation of nanofluids in the microfluidic T-junction and flow focusing geometries were conducted. Besides temperature-dependent droplet formation at both geometries, effects of other factors such as presence of

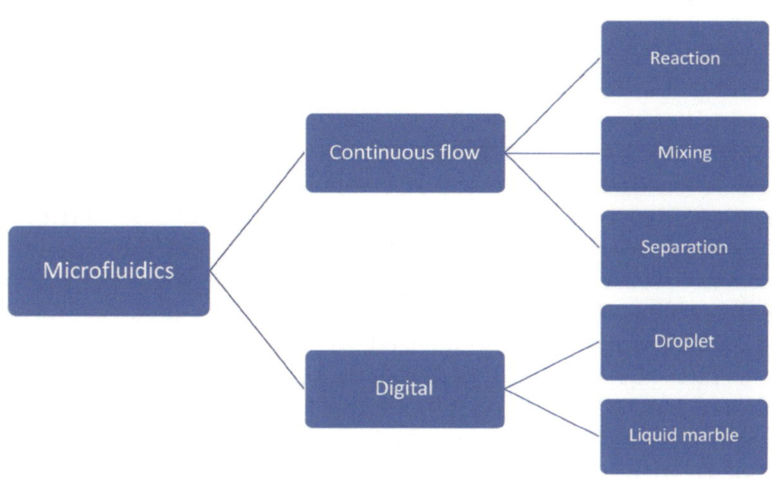

Figure 2.
Schematic of classification of microfluidic technology.

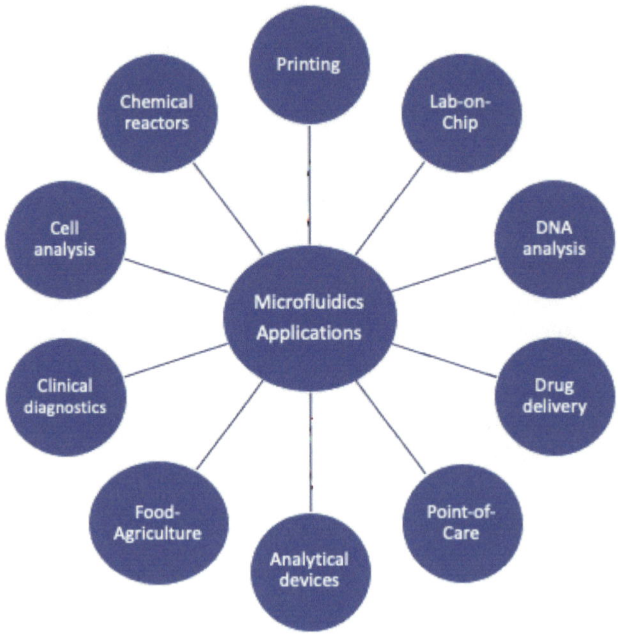

Figure 3.
Applications of microfluidic technology.

nanoparticles in aqueous fluid, depth of microchannel and flow rate on the droplet formation and size manipulation were investigated [10, 11, 16]. Although results are interesting and reveal the potential of nanofluids in microfluidics, more extensive research needs to be performed in this new combined field.

3. Overviewing nanofluids progress and challenges

The advancement in thermal management of modern equipment and systems are greatly impeded due to the limited cooling capabilities of conventional heat transfer fluids. Thus, there is an urgent need for heat transfer fluids with improved thermal properties and features. Nanofluids having found considerably higher heat transfer properties and features such as thermal conductivity, convective and boiling heat transfer compared to their base conventional fluids are believed to be capable of meeting such high cooling demands [8, 17–22]. With great features and properties nanofluids can be used for increased heat transfer and many other important fields of applications such as transportations (engine cooling or vehicle thermal management), microelectronics, solar energy technologies, micro-electromechanical systems (MEMS), electronics and instrumentations, heat exchangers, heating-ventilating and air-conditioning (HVAC), cooling electronics, microfluidics, defense, medical and so on (**Figure 4**). Researchers have also focused on studying nanofluids in diverse applications like- advanced cooling technologies, heat pipes, solar energy conversion and harvesting [8, 19, 23]. Being a multidisciplinary field and having numerous potential applications the impacts of nanofluids are very high. Although there are no recent market analysis or data for nanofluids, an estimation of the potential worldwide market for nanofluids was made sometimes ago and it was estimated to be over 2 B dollars per year only in heat transfer applications [21]. Majority of this market value came from cooling applications and nanoparticles (main component of nanofluids). Although extensive research works have been performed particularly during the past decade, nanofluids real applications and benefits are not yet achieved. Thus, more systematic and careful research works on nanofluids are still necessary.

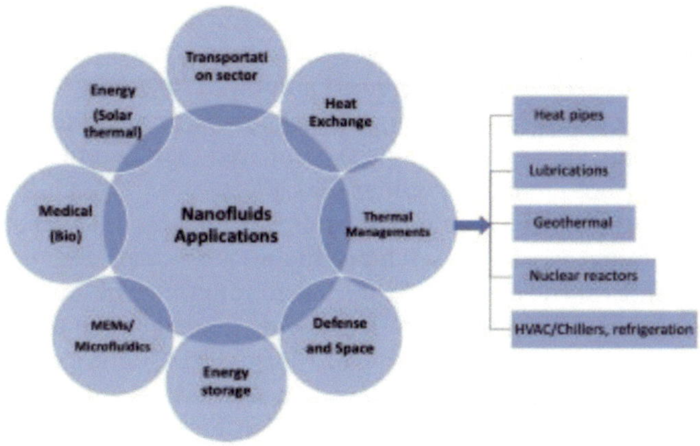

Figure 4.
Applications of nanofluids.

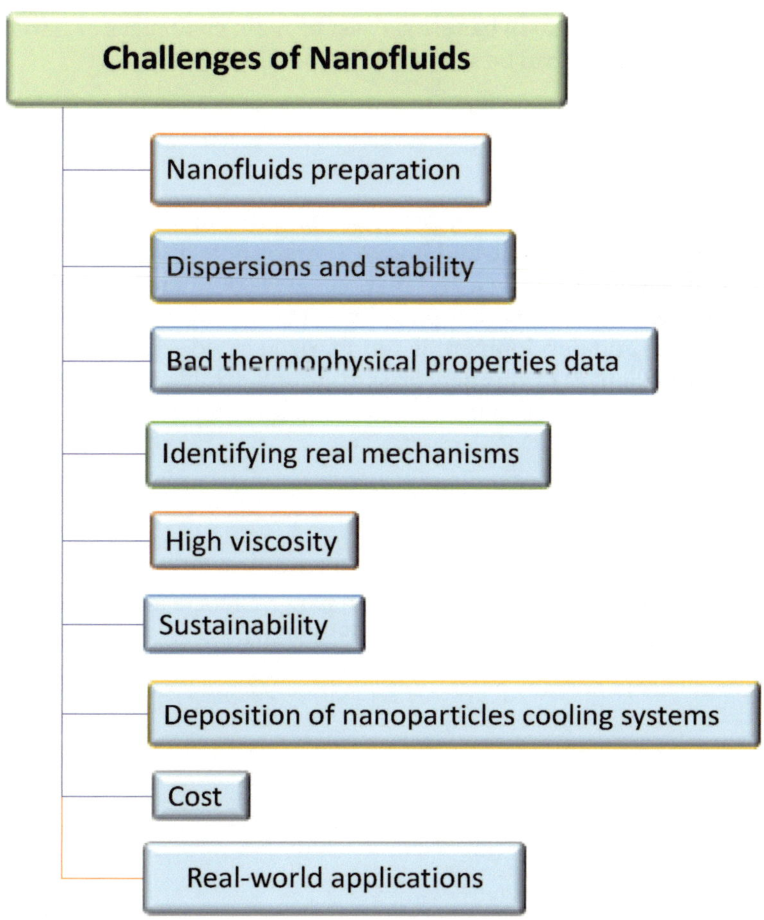

Figure 5.
Some key challenges of nanofluids.

Besides intensified research on various areas particularly on thermophysical properties and in several applications, research has been expanded to other new types of nanofluids like hybrid nanofluids and ionanocolloids (INC). Although research efforts on hybrid nanofluids have recently received increasing attention from researchers and findings from literature showed similar or better thermophysical properties compared to conventional nanofluids [24, 25], it is too early to explicitly identify their potential applications and benefits. On the other hand, ionanocolloids (suspensions of nanoparticles in ionic liquid) is another new type nanofluids which is in its early stage of research. However, ionanocolloids were found to exhibit superior thermal properties and show great potential compared to pure ionic liquids for many engineering applications including cooling [26–28].

Regarding research on thermophysical and heat transfer features, thermal conductivity is dominating the research producing more scattered data. However, the real mechanisms for the enhanced thermal conductivity of nanofluids are still inconclusive. Although many research efforts have been devoted until 2015 for the development of models for the prediction of thermal conductivity of nanofluids, the efforts have been declined recently despite no widely accepted models are available. This is due to the fact that no progress has been made on new mechanisms or physics-based understanding of underlying mechanisms. Nonetheless, despite above-mentioned impacts, potential applications, except few areas progressing on nanofluids towards developing them for real world applications are rather slow.

Nanofluids also possess serious challenges which impede their real progress towards application and human benefits. A list of main challenges of nanofluids research and development can be seen from **Figure 5**.

4. Remarks

This chapter gives an overview of progress, potential applications and main challenges of these two emerging and hugely important technologies. First it briefly discusses advancement of research and market values of microfluidics' followed by highlighting their main field of applications. Then, progress of and issues of nanofluids research and development is summarized besides bringing out potential application and challenges of this new class of popular fluids. This book is believed to be a useful reference source of information and knowledge in these two hugely important technologies.

Acknowledgements

This work was supported by the Fundação para a Ciência e a Tecnologia (FCT), Portugal through grant- PTDC/NAN-MAT/29989/2017.

DOI: http://dx.doi.org/10.5772/intechopen.98425

Author details

S.M. Sohel Murshed
Department of Mechanical Engineering, Center for Innovation, Technology and Policy Research (IN+), Instituto Superior Tecnico, University of Lisbon, Lisbon, Portugal

*Address all correspondence to: smurshed@tecnico.ulisboa.pt

References

[1] Manz A, Graber N, Widmer HáM. Miniaturized total chemical analysis systems: a novel concept for chemical sensing. Sensors and Cctuators B: Chemical. 1990; 1(1-6): 244-248.

[2] Song Y, Cheng D, Zhao L. Microfluidics: Fundamentals, Devices, and Applications. Weinheim: Wiley-VCH; 2018.

[3] Nguyen N-T., Wereley ST, Shaegh SAM. Fundamentals and Applications of Microfluidics. Massachusetts: Artech House; 2019.

[4] Masuda H, Ebata A, Teramae K, Hishinuma N. Alteration of thermal conductivity and viscosity of liquid by dispersing ultra-fine particles, Netsu Bussei. 1993; 7(2): 227-233.

[5] Choi SUS. Enhancing thermal conductivity of fluids with nanoparticles. ASME FED. 1995; 231:99-105.

[6] Das SK, Choi SUS, Yu W, Pradeep T. Nanofluids: Science and Technology. New Jersey: Wiley; 2008.

[7] Murshed SMS, Nieto de Castro CA. Nanofluids: Synthesis, Properties and Applications. New York: Nova Science Publishers. 2014.

[8] Murshed SMS, Leong KC, Yang C. Thermophysical and electrokinetic properties of nanofluids-a critical review. Applied Thermal Engineering. 2008; 28:2109-2125.

[9] Wen D, Lin G, Vafaei S, Zhang K. Review of nanofluids for heat transfer applications. Particuology 2009; 7: 141-150.

[10] Murshed SMS, Tan SH, Nguyen NT, Wong TN, Yobas L. Microdroplet formation of water and nanofluids in heat-induced microfluidic T-junction. Microfluidics and Nanofluidics. 2009; 6: 253-259.

[11] Tan SH, Murshed SMS, Nguyen NT, Wong TN, Yobas L. Thermally controlled droplet formation in flow focusing geometry: formation regimes and effect of nanoparticle suspension. Journal of Physics D: Applied Physics. 2008; 41:165501.

[12] Gravesen P, Branebjerg J, Jensen OS. Microfluidics-a review. Journal of Micromechanics and Microengineering. 1993; 3:168

[13] Convery N, Gadegaard N. 30 years of microfluidics. Micro and Nanoscale Engineering. 2019; 2: 76-91.

[14] http://www.yole.fr/Status_Of_The_Microfluidics_Industry_Market_Update.aspx (accessed on 10th May 2021)

[15] Ren C, Lee A. Droplet Microfluidics. London: Royal Society of Chemistry; 2020.

[16] Gupta A, Murshed SMS, Kumar R. Droplet formation and stability of flows in a microfluidic T-junction. Applied Physics Letters. 2009; 94:164107

[17] Murshed SMS, Nieto de Castro CA, Lourenço MJV, Lopes MLM, Santos FJV. A review of boiling and convective heat transfer with nanofluids. Renewable and Sustainable Energy Reviews. 2011; 15: 2342-2354.

[18] SMS Murshed, Nieto de Castro CA. Predicting the thermal conductivity of nanofluids- Effect of Brownian motion of nanoparticles. Journal of Nanofluids. 2012; 1:180-185.

[19] Murshed SMS, Nieto de Castro CA. Nanofluids as advanced coolants. In: Green Solvents I: Properties and

Applications in Chemistry, Eds. A Mohammad and Inamuddin, London: Springer; 2012.

[20] Murshed SMS, Nieto de Castro CA. Superior thermal features of carbon nanotubes based nanofluids- a review. Renewable and Sustainable Energy Reviews. 2014; 37: 155-167.

[21] Murshed SMS, Nieto de Castro CA. Conduction and convection heat transfer characteristics of ethylene glycol based nanofluids–a review. Applied Energy. 2016: 184: 681-695

[22] Cacua K, Buitrago-Sierra R, Herrera B, Pabón E, Murshed SMS. Nanofluids' stability effects on the thermal performance of heat pipes. Journal of Thermal Analysis and Calorimetry. 2019; 136;1597-1161.

[23] Khanafer K., Vafai K. A review on the applications of nanofluids in solar energy field. Renewable Energy. 2018; 123:398-406.

[24] Babu JAR, Kumar KK, Rao SS. State-of-art review on hybrid nanofluids. Renewable and Sustainable Energy Reviews. 2017;77: 551-565.

[25] Giwa SO, Sharifpur M, Ahmadi MH, Murshed SMS, Meyer JP. Experimental investigation on stability, viscosity, and electrical conductivity of water-based hybrid nanofluid of MWCNT-Fe2O3. Nanomaterials. 2021; 11:136.

[26] Nieto de Castro CA, Murshed SMS, Lourenço MJV, Santos FJV, Lopes MLM, França JMP. Enhanced thermal conductivity and heat capacity of carbon nanotubes-ionanofluids. International Journal of Thermal Sciences. 2012; 62:34-39.

[27] França JMP, Lourenço MJV, Murshed SMS, Pádua AAH, Nieto de Castro CA. Thermal conductivity of ionic liquids and ionanofluids and their feasibility as heat transfer fluids. Industrial & Engineering Chemistry Research. 2018; 57: 6516-6529.

[28] Minea AA, Murshed SMS. A review on development of ionic liquid based nanofluids and their heat transfer behavior. Renewable and Sustainable Energy Reviews. 2018: 91:584-599.

Chapter 2

Microfluidics for Time-Resolved Small-Angle X-Ray Scattering

Susanne Seibt and Timothy Ryan

Abstract

With the advent of new in situ structural characterisation techniques including X-ray scattering, there has been an increased interest in investigations of the reaction kinetics of nucleation and growth of nanoparticles as well as self-assembly processes. In this chapter, we discuss the applications of microfluidic devices specifically developed for the investigation of time resolved analysis of growth kinetics and structural evolution of nanoparticles and nanofibers. We focus on the design considerations required for spectrometry and SAXS analysis, the advantages of using a combination of SAXS and microfluidics for these measurements, and discuss in an applied fashion the use of these devices for time-resolved research.

Keywords: micro and nanoscale, systems design, lab-on-a-chip devices, SAXS, nanoparticles, time-resolved SAXS, microfluidics, hydrodynamic focusing

1. Introduction

Microfluidics is a multidisciplinary field dealing with the manipulation and behaviour of liquids and gases in dimensions below 1000 micron. The origin of microfluidics can be traced back to the 1970s, when miniaturisation became more and more developed. Applications in various fields, such as analytics, biology, chemistry, medicine and technology, became much more apparent with the development of *rapid prototyping*. Rapid prototyping describes a combination of photolithography, soft lithography and commercial printing, which makes the fast and efficient fabrication of custom designed microfluidic devices possible. Microfluidic devices for analysing aqueous samples were first introduced by Manz [1, 2], Harrison [3], Ramsey [4] and Mathies [5].

The most important benefit of microfluidic devices is their ability to perform quantitative and qualitative analysis with high sensitivity and resolution, while being a low cost method for fast, highly efficient analysis [6]. These factors make it especially useful for time resolved measurements, and coupling to small angle X-ray scattering (SAXS) measurements for the analysis of the average particle size and shape, and the evolution thereof under various *in situ* conditions. These approaches, in particular the coupling of microfluidics to SAXS, finds application in various areas, including biological materials, polymers, colloids, chemistry, nanocomposites, metals, minerals, food, pharmaceuticals and quality control [7].

Here we aim to detail background information important for the design of microfluidic devices for time resolved measurements, and the applications of these devices in time-resolved SAXS nanoparticle and self-assembly experiments.

1.1 Microfluidic principles

Fundamentally, the fluid dynamics in micro-dimensions are different from macroscopic systems. Fluid flows in these tiny systems are characterised by non-chaotic, smooth flow, where the fluid travels in parallel layers and the only interaction between those layers of flow is diffusion. By adapting reactions to microfluidic environments, the time axis of a reaction is converted into a distance axis along the outlet channel of the microfluidic device. This is key to enabling time-resolved studies *in situ* in a microfluidic channel.

The Navier–Stokes equation describes the motion of fluids mathematically, and is derived from Newton's second law of motion (F = ma), resulting in a set of two partial differential equations. For an incompressible Newtonian fluid, the Navier–Stokes equation is defined as:

$$\rho\left[\frac{\partial u}{\partial t} + (u \cdot \nabla)u\right] = \eta\nabla^2 u - \nabla p + F \tag{1}$$

where ρ is the density and η the viscosity of the fluid, p is the pressure, u the vector of the fluid flow, ∇ is the Nabla-Operator and F stands for any additional forces, that are directed at the fluid. The left side of the equation represents internal accelerations, and the right side represents the stress force per unit volume resulting from a pressure gradient and the viscosity of the fluid. In microfluidics, body forces are negligible, leading to a simplified, linear equation:

$$\eta\nabla^2 u = \nabla p \tag{2}$$

Treating the incompressible liquid as a continuum, the Navier–Stokes equation can be expressed as the continuity equation:

$$\nabla \cdot u = 0 \tag{3}$$

This means that the flux of liquid into a volume is the same as the flux out of a volume over a period of time. Additionally, the continuity equation is time-independent, restricting fluid flow in microfluidic channels to be symmetric in time [8].

To describe and compare phenomena on different scales, various dimensionless numbers for microfluidics were introduced. The most important is the Reynolds number (*Re*), showing the relation of inertial and viscous forces of a fluid. It is defined as:

$$Re \equiv \frac{\textit{inertial forces}}{\textit{viscous forces}} = \frac{|\rho\,(u \cdot \nabla)\,u|}{|\eta\nabla^2 u|} = \frac{\rho v}{\eta}d \tag{4}$$

where v is the flow velocity and d the characteristic length of the system, which in microfluidics is the diameter of the channel. The Reynolds number decreases with decreasing size of the system, reflecting the increased importance of viscous forces. The transition from turbulent to laminar flow is represented by Re being below 2040 ± 10.

The next most important dimensionless number is the Weber number (*We*), which describes the relation of the fluid surface tension to its internal forces, where γ is the surface tension of the fluid:

$$We \equiv \frac{\rho u^2 v}{\gamma}d \tag{5}$$

Microfluidic channel systems have generally a high surface-to-volume ratio, thus surface properties have significant effects on flow resistance and the velocity profile. To describe the interaction of a flowing liquid and a solid surface in microfluidic devices, Navier defined boundary conditions. The flow velocity tangential to the surface v_x is proportional to the shear stress at the surface and therefore given by:

$$v_x = \beta \frac{dv_x}{dy} \tag{6}$$

β is the slip length, or Navier length, and is defined as the distance from a point inside the channel to the surface, where the velocity is zero. Where $\beta = 0$ is a "no-slip" condition, describing the interaction between fluids and walls [9].

Every biological process or chemical reaction is limited by the converging and mixing of the reactants. Mixing in fluidic systems can generally occur *via* two methods – diffusion or advection. On the macroscopic scale, mixing is achieved by "chaotic advection" or turbulence, while on the micron-scale it is driven by diffusion. Diffusion specifies the migration of particles along a concentration gradient, and thereby always takes place from an area of high concentration to an area of lower concentration. This flux is in proportion to the diffusion coefficient, *D*, given by Fick's first law of diffusion. Solving Fick's diffusion law for adequate boundary conditions, the diffusion coefficient can be described for spherical particles with radius *r* in low *Re* numbers by the Stokes-Einstein relation:

$$D = \frac{k_B T}{6\pi\eta r} \tag{7}$$

with k_B as the Boltzmann constant, *T* as the temperature and η as the solvent viscosity. The relation between advection and diffusion for mass transport is described by the Péclet number (*Pe*) [8].

$$Pe \equiv \frac{advection}{diffusion} = \frac{vd}{D} \tag{8}$$

For turbulent mixing, advection dominates the above equation, leading to high *Pe* numbers. In microfluidics, turbulent chaotic mixing is very difficult to achieve, because the Reynolds numbers are almost always very low. Thus, in microfluidic channels, advection is almost always very small, and diffusion dominates, resulting in *Pe* numbers that are low. As such mixer design in microfluidics devices seeks to optimise diffusion [10, 11]. Along microfluidic channels, diffusion becomes insignificant when compared to convection occurring far downstream at the outlet channel. Thus most mixing devices incorporate some method for laminating flows to reduce diffusion distances, and reduce mixing times. Most commonly, these mixers are simple Y- or T-shaped cross channels, and diffusive mixing in these types of mixers for kinetic experiments can be described by the following equations:

$$\frac{d_1}{d_2} = \frac{\eta_1}{\eta_2}\frac{Q_1}{Q_2} \tag{9}$$

$$\frac{d_{MC}}{d_{SC_1} + d_{SC_2}} = \frac{\eta_{MC}}{\eta_{SC_1} + \eta_{SC_2}}\frac{Q_{MC}}{Q_{SC_1} + Q_{SC_2}} \tag{10}$$

where d is the thickness of the relevant layer, η the viscosity and *Q* the volume flow. The following assumptions must be fulfilled for these equations to be true:

1. The microchannel inhibits steady and laminar flow.

2. The fluids are all Newtonian.

3. Density and viscosity of all fluids is the same in all channels and do not change during the experiment.

4. The channel geometry is rectangular and all channel parts have the same height.

Eq. (9) applies to Y-shaped channel geometries where layer 1 and 2 are the spaces of two introduced liquid streams in the inlet channels, which merge in the outlet channel. For T-shaped channels where two side channels (SC_1 and SC_2) hydrodynamically focus a main channel (MC) stream, Eq. (10) applies [12].

1.2 Principles of small-angle X-ray scattering

Small angle X-ray Scattering (SAXS) is an extremely versatile technique used for investigating particle size, shape and dynamics that can be applied to a wide range of scientific problems. It is amenable to a wide range of particles, from the very small, of around a few nanometres, to very large sized structures in the order of a micron. It can be used to study mixtures, and the evolution of shape in reaction mixtures, and is widely used in biophysics and structural biology to confirm structure, and investigate structures that are not amenable to other structure investigations. SAXS can be used across all states of matter, including solids, liquids, gases, semisolid sample such as gels, and plasma. We will focus here on solution scattering, as this is the most applicable for microfluidic applications.

We aim to provide a brief overview of SAXS for solution scattering and time resolved measurements, but highly recommend Feigin and Svergun, 1987 [13] for a more comprehensive in depth review of SAXS measurements. In general, a solution SAXS experiment is relatively simple (which is one of the great attractions for the technique). A sample, in an appropriate sample cell, is exposed to a focussed, collimated monochromatic X-ray beam, and at a distance away from the sample the intensity of scattered X-rays is recorded using a 2D X-ray detector (**Figure 1B**). The resulting image is termed a scattering pattern. Similarly, the scattering from a matched pure background solvent is collected, and then subtracted from the sample scattering pattern to provide a scattering pattern that arises purely from the sample particles. The variation of the scattered intensity with angle, where the measured angles are very small, is related to differences in electron density between the sample and solvent, and the interatomic distances inside the sample particle, and thus contains information on the size and shape of the particle.

Scattering in solution is generally considered isotropic, as most particle systems adopt random orientations in solution. This allows for analytical mathematical descriptions of the scattering profile on the basis of particle shape. Scattered intensity (I) is described as a function of momentum transfer, q, and in a simplified form can be given as:

$$I(q) = \frac{N}{V} V^2 (\rho 1 - \rho 2)^2 P(q) S(q) \tag{11}$$

Where N is the concentration of the particle in the solution, V is the volume of the particle, $\rho 1 - \rho 2$ is the contrast in electron density between the solvent and the particle, and q is defined as:

$$q = 4\pi \frac{\sin\theta}{\lambda} \tag{12}$$

Where θ is the angle from the incident X-ray beam to the point on the detector where the intensity is measured, and λ is the wavelength of the incident X-rays (see **Figure 1A**). The derivation of the dependence of scattered intensity on the volume, concentration and electron density contrast of a particle described in Eq. 12 is given in detail in [13], which we highly recommend for further reading. The form factor $P(q)$ is typically a defined function, and varies depending on the physical parameters of the particle; for example a sphere with homogenous electron density has a different form factor function to that of a hollow sphere of the same size.

The structure factor component ($S(q)$) of Eq. (11) is a further analytical function that describes how the particles are arranged in the solution, e.g. forming large ordered structures with defined correlation lengths. Largely, samples are measured in a dilute condition, where the concentration of the particle is kept low enough to avoid these secondary interference effects, and thus $S(q)$ can be ignored. Where this effect cannot be avoided by reducing concentration, the use of hard sphere packing models or ionic charge–charge interaction models defining the effect as a function of q may be used to account for this effect, and provide information on changes in long range order in a sample.

Thus, for a sufficiently monodisperse sample, or a defined mixture of particles, it is possible to define an analytical model that provides volume, size and shape information. In polymer and colloid science, SAXS is used for many applications, including analysing the hierarchical nature of polymers in solution to assess clumping, local structure, overall morphology, and subunit arrangement, assessing the shape, size and dispersity of nanoparticles in solution, and investigating the dynamics, and evolution of particle size and shape under varying solution

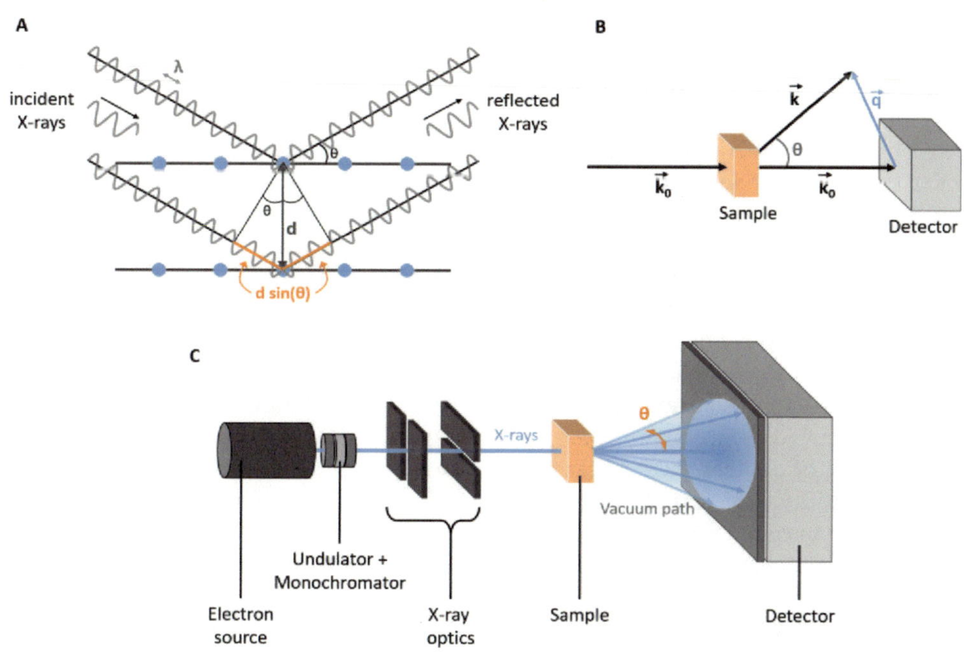

Figure 1.
(A) Schematic illustration of the Bragg equation with incident and reflected X-rays on two scattering planes, showing the lattice distance d, the half scattering angle θ, the wavelength λ and the path difference defined by Bragg's law. (B) Geometric construction of the scattering vector q from the incident wave vector k_o and the scattered wave vector k with the half scattering angle θ. (C) Schematic setup of a small-angle X-ray scattering setup.

conditions and chemical reactions. SAXS is clearly a versatile technique that can provide useful information on systems that are well behaved, and can also be applied to samples that may not display ideal behaviour (for example aggregation prone nanoparticles, or time dependent mixtures of particles). However, the measurement does have some drawbacks. SAXS analyses are heavily reliant on complementary information. SAXS cannot provide information at an atomic resolution, so high resolution structural information is lacking, and needs to be obtained by alternative methods such as NMR, chemical crystallography, or electron microscopy. Further, SAXS does not provide information on changes in chemical environment so correlating the particle shape and size evolution with changes in the chemistry of a system requires the use of other techniques that are sensitive to the chemical environment. Additionally, for *in situ* experiments, SAXS on high intensity beamlines has the disadvantage that intense dose of radiation are required to obtain high quality data at short time frames. This can result in radiation damage in the sample that can significantly influence results.

1.3 Microfluidic devices and X-rays

In SAXS analyses, there are a range of disadvantages that the current sample environments struggle to address. First and foremost is that in most solution SAXS measurements there needs to be a high concentration of particles in the solution to achieve a scattering signal with high enough signal to noise to be of use in further analysis. For the most part, this is not a significant issue as most samples are generally amenable to reasonably high concentrations. However, in a number of cases, the amount of sample can prohibit the use of standard sample environments, and limits the use of SAXS to samples that are not in limited quantities, or expensive to produce. Further, for a continuous flow mixing device, where many exposures are required at each time point, the sample consumption can reach many millilitres; again this may be prohibitive for a majority of samples. Additionally it can be difficult to apply high throughput methodologies to systems where flow, volume and data quality constraints limit the number of measurements that can physically be conducted in a period of time.

The limitations of the current sample environments can be significantly mitigated by the use of custom microfluidic devices. The very low internal volumes mean that sample consumption is reduced, and the time that a volume of sample can be measured over under flow is increased, leading to a general improvement in measurement statistics. The lower spatial footprint, and lower sample consumption rates, means that a large number of measurements can be conducted in a very short period of time in parallel; increasing throughput for screening measurements. The lower volumes, and thus much more efficient mixing allows for much lower deadtimes then would otherwise be possible, and with the increasing access to microbeam SAXS measurements, the time resolution of the mixing experiments are greatly improved over conventional approaches. Further, the ease of design and modification of devices means that bespoke devices for specific applications can be achieved rapidly. Given that microfluidic devices can address many of the limitations of conventional SAXS sample environments, we believe that there will be increasing uptake and incorporation of these devices into SAXS measurements.

2. Microfluidics for time-resolved studies

2.1 Device design

To successfully investigate time-resolved reactions in microfluidic devices, the channel design has to be carefully adapted to the requirements of each application.

All experiments require planning and consideration of the simultaneous use of analysis, mixing and cleaning equipment due to the generally small dimensions of microfluidic devices. In the past decades, a very diverse range of microfluidic reactor devices have been designed for time-resolved studies of reactions. Designs such as continuous-flow, stopped-flow, droplet-based and digital microfluidics have been developed and applied to produce materials with sizes ranging from nanometres to almost millimetres. In this chapter, we are focusing on continuous and stopped flow devices, in particular on hydrodynamic focusing techniques. In comparison to droplet-based techniques, hydrodynamic focusing is a straight-forward approach to implement, due to its pure hydrodynamic principles. It only includes surface tension effects at the liquid–liquid interface in the outlet channel of the microfluidic device without the need of consideration of surface tension effects at liquid–gas interfaces. These devices offer stability at high flowrates, allow high-throughput applications and enable highly controllable operational conditions, as the flow behaviour is the only influential parameter that needs to be considered for time-resolved studies.

2.1.1 Flow field considerations

An understanding of flow fields at the microscale is required to understand the function of hydrodynamic focusing and device design considerations. No turbulent mixing occurs inside a microfluidic channel, as typically *Re* numbers below 100 are achieved, thus liquids can only mix by diffusion. This has the advantage of allowing predictions of the exact movement of particles by calculation, as no chaotic (turbulent) mixing needs to be considered.

For microfluidic channels, assuming no-slip conditions in combination with pressure driven flow, Poiseuille flow with a parabolic shaped flow profile arises. Here, the highest velocity is in the middle of the channel, which decreases parabolically towards the walls until it reaches zero. For cylindrical shaped channel geometries with coordinate length x, radius r and azimuthal angle Φ, the velocity field can be derived as:

$$\nu_x(r, \phi) = -\frac{\Delta p}{4\eta L}\left(a^2 - r^2\right) \tag{13}$$

With pressure p and viscosity η over the channel length L and channel radius a. The hydraulic resistance R results then as:

$$R = \frac{8\eta L}{\pi a^4} \tag{14}$$

For rectangular shaped channels with height h, width w and small aspect ratio $(w > h)$ the velocity field over the coordinates x, y, z is:

$$\nu_x = \frac{4h^2 \Delta p}{\pi^3 \eta L} \sum_{n,odd}^{\infty} \frac{1}{n^3} \left[1 - \frac{cosh\left(n\pi \frac{y}{h}\right)}{cosh\left(n\pi \frac{w}{2h}\right)}\right] sin\left(n\pi \frac{z}{h}\right) \tag{15}$$

and the hydraulic resistance R is then [14]:

$$R = \frac{12\eta L}{wh^3}\left[1 - \frac{h}{w}\left(\frac{192}{\pi^5} \sum_{n=1,3,5}^{\infty} \frac{1}{n^5} tanh\left(\frac{n\pi w}{2h}\right)\right)\right]^{-1} \tag{16}$$

This understanding of this pressure-driven, steady-state flow in microfluidic channels is the basis of liquid handling in lab-on-chip systems. Especially in microfluidics, the channel cross-sections can be of various shapes, depending on the application and fabrication method. Eqs. (13) to (16) describe the velocity field and hydraulic resistance for spherical and rectangular cross-sections, which are the most common geometries used for the devices described in this chapter. The derivation of those values is exceedingly more complicated for arbitrary channel cross-section shapes.

2.1.2 Continuous flow vs. stopped flow

The first design consideration is how the device will enable time-resolved measurement of phenomenon. This can be achieved in two different ways. The first, and conceptually simplest method, is a static experiment, where a sample is firstly mixed and then introduced into a monitoring chamber and measured repeatedly at defined time periods. The most common apparatus for this style of measurement this is a *stopped flow* device, where mixing is achieved rapidly, and then flow is stopped as soon as the homogenously mixed sample fills the monitoring chamber. The measurement is triggered as soon as the flow is stopped, and generally continues as rapidly as possible until the reaction reaches completion. The second method is to use a continuous flow system, where the mixed sample is introduced into a flow-through system, and temporal measurements are achieved by varying the distance between the mixing point and the sampling point.

Both styles of devices have advantages and disadvantages, and the choice depends strongly on several experimental considerations, including the time domain of the reaction, mixing efficiency, sample volume constraints, and sample chemistry constraints (e.g. resistance to photobleaching, or radiation damage). Stopped flow measurements are favoured when there is a small volume of sample that is resistant to measurement induced damage (for example a flurophore that is resistant to photobleaching), where the reaction is not extremely fast, and where the experimental measurements are not slow. In stopped flow measurements the initial point in the measurement is always some degree of time post the start of the reaction (given the time it takes to fill the sample cell, stop the flow and take the first measurement), and the temporal resolution of the measurement is given by the speed at which the measurement can be taken. However, agglomeration of the reacting sample on the channel walls can influence the quality of measurements and, due to the ongoing reaction, leads to only a small window that can be detected before the experiment needs to be repeated. Alternatively, continuous flow measurements favour samples that are sensitive to the measurement, are very rapid, and require temporal resolution finer then the measurement speed of the instrument. Continuous flow measurements allow for measurement very close to the point of mixing, temporal resolution is given by the spatial resolution of the measurement, and the time taken to travel to the point of measurement. Further, the deadtime and temporal resolution is heavily influenced by flowrate, allowing for fine control across many temporal regions. As a result, the observation of the reaction can be precisely controlled. It needs to be considered that continuous flow measurements need more sample volume in comparison to stopped flow methods, to provide a constant flow profile.

2.1.3 Hydrodynamic focusing

The basis of hydrodynamic focusing lies in a central solution that flows with a lower flowrate within an outer sheath fluid with a higher flowrate. This enables the compression, or focusing, of the central flow, and decreases the mixing times significantly by reducing the diffusion length. There are two main categories for microfluidic devices with hydrodynamic focusing: coaxial tube and planar on-chip devices. The design can be defined depending on the use, with adapted geometries for fast mixing (**Figure 2A**), gradients (**Figure 2B**), specific nanoparticle growth reactions or self-assembly processes (**Figure 2C**) [15].

The simplest type of coaxial tube reactors is a device consisting of two concentric capillaries (**Figure 2D**), which are connected to a channel where a central flow is injected through the inner capillary, with sheath flow injected from the outer layer. Coaxial tube microreactors find various applications, but are typically used as the interface for droplet-based reactors, as the transition from flow to droplet generation, dripping and jetting is defined by the flowrate of the outer sheath flow. However, this approach is of limited utility in time resolved mixing applications, as it only offers limited mixing geometries. Further limitations abound in the fabrication process for coaxial designs, which requires multiple steps and precise alignment and assembly of the parts [15].

On-chip hydrodynamic focusing devices can be differentiated in two-dimensional (2D) and three-dimensional (3D) devices. In two-dimensional hydrodynamic focusing devices, the central flow is focused in the horizontal plane. The simplest geometry for a 2D device is a Y- or T-shaped mixer (**Figure 2E**), where the cross-sectional diffusion is broadened at the channel walls in comparison to the centre. However, this design is highly limited regarding flow stability and focusing and susceptible to variation in these parameters in operation. To avoid this, cross-

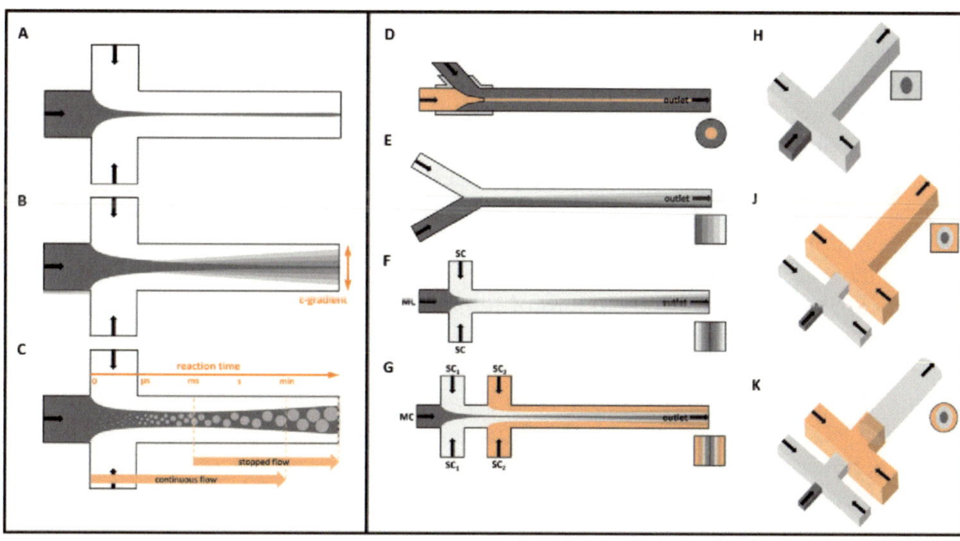

Figure 2.
Left: Scheme of microfluidic features for kinetic investigations in flow in a cross shaped mixer. (A) Hydrodynamically focused Centre stream for fast mixing experiments. (B) Mapping of concentration gradient across and along the channel through interdiffusion of different liquids from main and side channels. (C) Nucleation and growth of nanoparticles or self-assembly processes of nanomaterials as a function of time along the outlet channel. Schematic comparison of the provided time scales in continuous and stopped flow microfluidic devices. Right: Schematic illustration of microfluidic devices with various channel cross designs with the corresponding cross-sections through the outlet channel. (D) Coaxial tube reactor with two concentric channels/capillaries. (E) Y-shaped design, where mixing is solely based on diffusion. (F) Cross-shaped geometry at the inlets for hydrodynamic focusing. (G) Two-cross-section geometry, also known as double-focus device, where three different solutions can be introduced into the channels. Solutions introduced into the first side-channel (SC_1) act as an inert buffer between reactants in main channel (MC) and second side-channel (SC_2). (H) and (J) multilayer designs of the geometries from (F) and (G), respectively, avoiding contact between the central stream and channel walls. (K) Hybrid device consisting of multilayer focusing device (J) and an inserted glass capillary as outlet channel.

shaped geometries can be employed, where the central flow is focused from both incoming side channels (**Figure 2F**). This provides good control over the thickness of the central stream. Furthermore, it allows a well-defined sample composition that can be adapted by variation of the volume flow in each inlet individually. To adapt the device design to multi-step synthesis, several side channels can be added to introduce additional reactants (**Figure 2G**). Thus planar on-chip hydrodynamic focusing is more highly favoured for flexible mixing devices, and is used much more often in general microfluidic designs.

However, there are further considerations that need to be taken into account, particularly for chemical reactions or self-assembly processes. In coaxial and 2D channel geometries the interface of reacting solutions is in contact with the channel walls, and particles or macromolecules can stick and agglomerate on the channel surface and disturb the laminar flow conditions. Furthermore, this accumulation interferes with analytical investigations and, in the worst case, can cause complete blockage of the channel.

To avoid channel contact, three-dimensional channel geometries can be used. 3D hydrodynamic focusing requires both horizontal and vertical focusing of the central reactant stream, leading to a complete enclosure with liquid from all sides (**Figure 2H** and **J**). The device design can be optimised to reach homogeneous mixing without integrating specific mixing regions before the measurement part of the microfluidic chip. Additionally, these devices are simple to fabricate and have easily adjustable designs. The most common design to achieve 3D hydrodynamic focusing is through multi-layer on-chip devices, which require precise alignment as part of the fabrication process. Alternate methods are single layer devices or novel fluid manipulation technologies like "microfluidic drifting", which introduces lateral drifts or counter-rotating vortex forces to achieve vertical and horizontal flow focusing. These alternatives require less alignment in manufacturing and are thus much more user friendly in regards to fabrication.

All previously described device geometries can be used to produce droplets, liquid jets and sprays under the right flow conditions, including ultra-high flowrates. These devices are not limited to constrained flows inside channels, and for time-dependent studies the use of free liquid jets is preferred. Measurements of free jets have significant advantages in many optical measurements, as there is little to no background signal from surrounding material. When employing free jet devices, the parabolic flow profile from laminar flow within channels turns into a plug flow profile after passing the nozzle outlet. The liquid–solid interface of the no-slip condition resulting in parabolic flow is replaced by the liquid–gas interface in air, which has lower friction with the fluid and can be accelerated in flow direction. Free jets, however, are also quite difficult to work with, and despite the advantages in background have not found widespread use in the field.

2.2 Fabrication

2.2.1 Fabrication techniques

Many different technologies exist for the production of microfluidic devices. In general, fabrication of microfluidic devices in hard materials is often very time-consuming and cost-intensive, thus polymers are generally preferred, particularly when cost and ease of fabrication are considerations in the design process. In most cases, designs are started in silicon, with a Computer-Aided Design (CAD) model of the device. The device is then fabricated using any of a number of different technologies following a process of rapid prototyping. We focus here on some of the more common approaches.

Lithography. One of the most powerful methods in microfabrication is lithography. It can be differentiated by the type of radiation used, e.g. photolithography, electron-beam lithography, or X-ray lithography. With these different lithography methods, structures with sizes between 0.2 and 500 μm in hard materials like glass, or between 0.5 and 500 μm in soft materials like polymers, can be achieved. In the most common form of lithographic fabrication, a UV blocking mask is generated from the CAD model, and adhered to a silicon wafer. The master model for fabrication is then generated by photolithography, where the masked wafer is coated in a photosensitive epoxy monomer solution, and UV-cured. In general these silicon masters are then used to generate working devices via soft lithographic replication. The most common approach is to use the master chip as a mould, and poly (dimethylsiloxane) (PDMS) to form an imprinted device, which can be bonded to a glass-slide or a second PDMS part to form the final microfluidic device [16–18]. The replicated structure can be a positive or negative of the initial design, depending on which part of the structure was UV-cured on the silicon master.

Hot embossing and micromoulding. Hot embossing is a micromoulding technique that uses thermoplastic polymers to imprint structures at elevated temperatures. It usually uses high-temperature polymers, e.g. PMMA, PC, PI, PE, PVC or PEEK, which are heated above their glass transition temperature (T_g) before being pressed into a mould with high pressure. The moulds have to withstand the applied pressure and high temperatures, and are often made of metal or silicon, fabricated via etching, lithography in combination with electroforming and moulding or CNC (Computerised Numerical Control)-machining. The accuracy of hot embossing is in the order of tens of nanometres, making it possible to obtain high aspect ratios of structure, while being a low cost and easy procedure. It is often used with a defined and tested device design as high throughput method with a very short fabrication time [19, 20].

3D printing. Within the last decade, 3D printing technologies have advanced to astonishing precision, in size-regimes down to the micrometre scale. Additive manufacturing technologies like fused deposition modelling (FDM), stereolithography (SLA) or selective laser sintering (SLS), have been developed for various materials like polymers, resins, ceramics or metals. It is possible with these techniques to produce a complete microfluidic device in one step. The device material and process can thereby be selected with regard to required mechanical and chemical properties of the device. A channel size resolution of few hundred micrometres can be achieved, making this approach preferred for devices with wider channels [21, 22].

2.2.2 Device materials

Based on the desired purpose of the microfluidic chip, device materials must fulfil specific criteria. The most important requirements which will be addressed in this chapter are solvent stability, ease of fabrication, and optical and X-ray transparency.

Solvent stability. Microfluidics deals with the manipulation of liquids, which means that the device material has to be resistant and inert to the solvent. This becomes especially relevant when using organic solvents, as they often cause swelling or dissolution of standard polymeric device materials. Swelling leads to deformation, which can cause channel closure. A number of device materials have been tested with regard to resistance to some common solvents for nanoparticle synthesis and self-assembly processes (**Table 1**). It is clear from these results that careful selection of polymer is necessary for long term stability.

Ease of fabrication. While the solvent is important, it is also essential to consider the difficulty of working with the various polymers, and the end

Polymer Solvent	PDMS	NOA 81	THV815 GZ	THV610 AZ	THV500 GZ	THV221	ET 6235	HTE-1705Z	PS	PMMA	TOPAS 8007/6013	Bendlay	SIFEL	SU8 50
THF	x	x	x	o	o	x	o	(x)	x	x	x	x	o	o
Toluene	x	(x)	x	o	o	o	o	o	x	o	x	x	o	o
Chloroform	x	x	x	o	o	o	o	o	x	x	x	x	o	o
Dioxane	(x)	x	x	o	o	o	o	o	x	x	o	x	o	o
Acetone	o	o	x	x	o	x	o	o	x	x	o	x	o	—
Octadecene	o	o	x	o	o	o	o	o	o	o	x	o	—	—

x: dissolution or swelling of the material, o: no change observed, –: not tested. Results in () showed insignificant swelling, the device could be continued to be used [23].

Table 1.
Stability test of various polymers and the corresponding solvents.

characteristics of the device. For example, NOA 81 is a turbid, commercially available, UV curable polymer mixture from Norland Optical Adhesive, which is relatively easy to work with. However, devices made from NOA 81 are thin and relatively flexible, even after sealing top and bottom half, so it should be avoided if stiff or thick devices are required. In comparison, SIFEL (SIFEL2610) is a fluorinated polymer distributed by Shin-Etsu that is liquid at room temperature and hardens at higher temperature, and is stable against all tested organic solvents. The device fabrication however, is time consuming, requiring the additional step of sputtering the silicon wafer with an inert chemical layer to allow release of the SIFEL device from the mould.

Materials can also dictate the method of fabrication, for example THVs (fluorothermoplastics of blended tetrafluoro ethylene, hexafluoro propylene and vinylidene) must be fabricated by hot embossing. Glass or hard material devices are made with difficult fabrication techniques, like etching. In many cases prototypes are made with cheap, easy to fabricate materials, with the more difficult fabrication only for the final working devices where needed.

Optical and X-ray transparency. The most commonly used method for alignment of device parts and analysis of ongoing reactions is optical microscopy. Hence, the optical properties, e.g. transparency, of the microfluidic devices should be considered. Further, the final measurement modality must be considered in material selection. For example, if three-dimensional confocal microscopic investigations of the whole channel volume are required, the selected device material should provide a low absorption behaviour in the range of the sample-specific selected laser wavelength, and low fluorescence background. Or as the focus of this chapter is SAXS, the material of the device should have high X-ray transmission, and low scattering in the q-range of interest. The material should also be able to withstand the X-ray radiation, which is present on high flux SAXS beamlines. In our experience, the lowest background scattering for higher q measurements above 0.05 $Å^{-1}$ were achieved with glass, NOA81, PDMS and Kapton. Other polymers such as THV and TOPAS showed diffraction and correlation peaks in the high q region >0.1 $Å^{-1}$, that interfere with background subtraction, and worsen signal to noise. For measurements at low scattering angles with q values under 0.05 $Å^{-1}$, glass, PMMA, PS, NOA81 and TOPAS display flat scattering curves. All other tested materials at this q-range showed significant scattering signals from the device [23]. Furthermore, although showing a low scattering background at high scattering vectors, PDMS was extremely sensitive to radiation, deforming the channel and showing an increasing and changing scattering profile with exposure. This material is typically unsuitable for SAXS measurements.

Hybrid microfluidic devices can marry the best characteristics of materials, to achieve a successful device. For example, the complex mixing cross section can be made from easy to handle materials, e.g. PDMS, and a robust X-ray transparent, low background scattering material inserted as an outlet channel after the last cross-section, e.g. a glass capillary (**Figure 2K**). These devices have the advantage of high optical and X-ray transparency in the measurement region, while allowing adjustment to the mixing cross design in the polymer part [23, 24].

2.3 Practical considerations for device handling

2.3.1 Fluid handling

A key practical consideration for the use of microfluidics is the method for introducing fluid into the device. For the most part, each interface channel should have its own fluid handling system, which should be capable of smooth, pulse free flow, with no bubbles or leaks, and should have similar chemical compatibility to that of the microfluidic device [25]. We favour modular syringe pump systems, which have the ability to adapt the amount of dosing units to the number of channels. Other options include flow regulated gear pumps, positive air pressure systems or even on-chip fluid reservoirs. In any case, fluid flows should be accurately calibrated immediately prior to use, to ensure the correct dosage, flowrates and thereby the correct flow profile. A further consideration in fluid handling is minimising dead volumes (in particular by using appropriate fluidic connections and minimising tubing lengths), to prevent wastage of sample. Further, it is ideal, particularly for time resolved SAXS experiments where access to the system is restricted, if the fluid handling system can be controlled and triggered remotely, as this allows for accurate initiation of the reaction and data acquisition. The usability of all devices should be tested before each experiment to avoid leakage and proper function of the channels, especially with regard to flow focusing. Tubing and device failure are common frustrations in obtaining good data.

2.3.2 Temperature control

In general, homogeneous temperature control of the reaction solution has to be achieved. It is possible to submerge the whole microfluidic device and tubing in a water / oil bath. However, for *in situ* investigation, there needs to be unimpeded access to the channel, and this approach is thus not viable. In this case, custom designed heating elements, e.g. heated enclosures, are employed to regulate temperature. Temperature control is only limited by the geometric constraints of the measurement, and the heat transmission of the device material. For example, we have implemented copper heating tubes for surrounding the glass capillary of hybrid microfluidic chips, incorporating a window for the X-ray beam that provided excellent thermal control of the measurement [26, 27]. A key to good thermal stability is to also incorporate heating elements for the fluidics systems, to keep the reaction solutions at appropriate temperatures and ease the thermal load on heating elements in the device.

3. Microfluidics and SAXS

3.1 Incorporating microfluidic experiments into SAXS

After the microfluidic devices are designed, fabricated, tested and fluid control is established, final considerations involve the implementation of the complete setup in an X-ray beam, either in a SAXS lab instrument or at a synchrotron beamline. Depending on SAXS instrument design a number of adjustments and considerations are required to achieve good integration for the measurement. As every synchrotron has slightly different parameters and sample environments, it is recommended to contact the beamline staff if considering a microfluidic-based time-resolved SAXS experiment for specific advice.

3.1.1 Device modifications

SAXS measurements are dependent on the volume and composition of all objects in the X-ray beam. Consideration should therefore be given to not only the material's resistance to radiation damage, but also the relative volumes of device material and sample that are going to be presented over the measurement channel. For example, 2–3 mm of any device-polymer either side of a sample channel of 50 micron means that the scattering from the sample in the channel will be entirely masked by the scattering of the polymer device, even if the polymer scattering is low. Further, the more material that is in the beam pathlength the more attenuation of the X-ray beam will occur. This means that less photons will hit the sample, get scattered, and escape the device to be detected. Calculations based on composition and thickness of the material should be done in advance to determine the expected transmission of the device. In many cases, this requires a redesign of the device itself to thin down the supporting material around the channels, incorporate X-ray transparent windows, or change the device material.

3.1.2 Device mounting in beamline/SAXS instrument

It should be expected that the device will be required to be perpendicular to the beam. Further consideration should also be given to the orientation of the channels, with respect to the beam dimensions. Generally, it is optimal to orient the channels so that as much of the beam is going through the channel as possible, and as little as possible is hitting the device body. This minimises background, and optimises the signal that can be achieved. In the best case scenario, the beamline will have the capability to generate micro beams of a few micron in any dimension. This allows for optimal exposure for the sample, and greatly increased time resolution in time-resolved samples.

It is best if the device has a chip-holder to mount the device in, which in most cases is specific to the setup and design. This holder must allow for any necessary connections of inlet and outlet tubing while holding the microfluidic device steady and without tension on any connections to pumps or vials. Ideally, this holder would be placed on a motor-controlled, adjustable stage to facilitate precise alignment in the X-ray beam and movement of the device to scan along outlet channels for different points in time of reaction kinetics.

In many cases, beamlines and lab instruments will maintain a vacuum along the complete X-ray flight path, and may include a vacuum sample environment. As X-rays interact with all matter, it is a requirement that there not be air in the majority of the SAXS instrument. Vacuum sample environments take this further by removing all air in the system to reduce and minimise background scattering. If a vacuum sample environment is in use, the microfluidic device must be designed to withstand the vacuum levels, and to minimise outgassing and other deleterious effects.

3.2 Nanoparticle nucleation and growth

A fundamental principle in nature and technology is self-assembly – the formation of ordered structures of components of a system out of chaotic arrangements without external forces. These processes can be induced by a multitude of parameters, e.g. change of solvent, pH, temperature, pressure or by introduction of additional reactants. SAXS, being sensitive to length scales of 1–100 nm is an ideal technique for studying nanoparticle size and structure from nucleation to the final particle. *In situ* SAXS measurements of nanoparticle synthesis is typically used to monitor the kinetics of this process [7], and increasingly incorporates microfluidics.

Metal nanocrystals. Metal nanoparticle syntheses is particularly amenable to SAXS analysis, as their high electron density contrast allows measurements in dilute suspensions even at the very early stages of particle nucleation. This has been employed for investigating silver (Ag) and gold (Au) nanoparticle formation and structure [28–31]. The first steps towards microfluidic setups were stopped flow measurements, for example the kinetics of gold nanoparticle formation, and the concurrent evolution of the optical properties of the particle at room and high temperature was very successfully investigated at millisecond resolution with this method by Abécassis et al. [32, 33] and Chen et al. [26]. Further development by Polte et al. provided *in situ* studies on the nucleation and growth of Au and Ag nanoparticles in stopped and continuous flow microfluidic devices [34, 35] Free liquid jets coupled to microfluidic mixers have aided in reducing background and improving signal to noise of SAXS measurements [36].

When combining microfluidic setups and X-ray scattering for nanoparticle investigation, not only the reaction kinetics of nucleation and growth processes can be measured, but also agglomeration kinetics and structures. For example, Gerstner et al. combined a static microfluidic mixing device with in line absorption and SAXS measurements to study the rapid superlattice formation of alkylthiol-coated Au nanoparticles at different temperatures, which showed a differentiation between long- and short-range self-assembly effects of temperature on a time scale down to 3 seconds [37]. A further example is the time-resolved analysis of polystyrene (PS)-coated Au nanoparticles by Merkens et al. in a Kapton-based 3D hydrodynamic focusing microfluidic chip, that revealed the subsecond kinetics of structural transitions involved in solvent induced collapse [38].

Semiconductor nanocrystals. Inorganic semiconductor nanoparticles, also called quantum dots (QDs), have received much attention due to their bright and size-tunable photoluminescence, which is commonly used as a key measurement property during synthesis [39]. During a synthesis of QDs, inorganic particles undergo a process of nucleation, growth and agglomeration, followed by dispersion into a buffer solution to quench the reaction. In order to synthesise homogeneous particles it is important to induce rapid nucleation and control the growth rate. Microfluidic devices with hydrodynamic focusing have been extremely useful in achieving this controlled synthesis process [15]. We have used 3D hydrodynamic focussing device for the synthesis of CdS nanoparticles, both studying the reaction by confocal laser scanning microscopy (CLSM) and SAXS. The CLSM measurements, using a full-PDMS device, showed the increase and shift of photoluminescence related to the nucleation and growth of CdS nanoparticles along the outlet channels. Hybrid microfluidic chips, consisting of the mixing cross section in PDMS and an inserted glass capillary as outlet channel (**Figure 3D-F** D-F), were developed for *in situ* SAXS measurements with low scattering background [24]. Employing a stopped flow setup, the nucleation and growth of ZnO nanoparticles was characterised at the timescale of seconds [27]. Further work elucidated the kinetics of the process at the microsecond timescale, using a free-jet device with a microfluidic T-mixer setup with a nozzle outlet to perform synchrotron SAXS measurements of the reaction in air (in the free jet). These setup enabled the investigation of QD synthesis with and without stabilising agents [42], highlighting the use of microfluidics and SAXS in the development of straightforward processes for nanoparticle synthesis.

3.3 Macromolecular self-assembly

Structural evolutions of pure and mixtures of surfactants that are often used in nanoparticle synthesis reactions, can also be investigated by a combination of microfluidic platforms and SAXS. Fürst et al. used a simple, T-shaped microfluidic chip to measure the structural assembly of tetradecyldimethylamine oxide (TDMAO) and lithium perfluorooctanoate (LPFO) in combination with synchrotron SAXS. This revealed the kinetic fusion mechanism of the cylindrical TDMAO and spherical LPFO micelles to disk-like micelles as a diffusion limited process, resulting in lamellar correlations at final stages [23].

Amphiphilic diblock copolymers show fast self-assembly processes at a timescale of seconds. These can be followed *in situ* with specially designed equipment by synchrotron-based SAXS, as shown by Stegelmeier et al. for PS-P4VP block copolymers by rapid removal of solvent [43]. An elegant way to study these fast self-assembly processes *in situ* in solution is shown by With et al. by measuring the concentration-induced lyotropic phase transition of PI-PEO polymers. Employing a simple cross-shaped multilayer Kapton microfluidic device (**Figure 3A-C**) in combination with synchrotron microfocus SAXS, time-resolved self-assembly of the used PI-PEO polymers via a spinodal microphase to micelles into FCC liquid-crystalline phases could be studied with millisecond resolution [40]. A more sophisticated channel design was used to study the self-assembly of PI-PEO block copolymers via spherical micelles into a FCC lattice (**Figure 3G-J**) and the solvent-induced self-assembly of PEG-PLAinto spherical micelles, cylindrical micelles and vesicles by Fürst et al. [23].

Apart from surfactants, polymers and polymer coated particles, other materials have self-assembly properties and can be investigated with a combination of microfluidics and X-ray scattering. Seibt et al. followed the pH-induced, rapid assembly of disk-shaped hydrogelators (N,N′,N″-tris(4-carboxyphenylene)-1,3,5-benzene tricarboxxamide) to nanofibrils of several hundred nanometres length by CLSM and SAXS. The measurement of the self-assembly process utilised 3D hydrodynamic focusing microfluidic devices (**Figure 3D-F**). Even bigger structures could be followed in the case of collagen and collagen derived fibres by pH-induced self-assembly. Here microfluidic chips provide an excellent platform for wet-spinning processes, shown by Haynl et al. [44] and Hofmann et al. [45], while SAXS can provide important information about the internal structure of the fibres during formation [46]. Furthermore, the alignment of macromolecular structures, such as worm-like micelles, patchy polymers and nanoplatelets can be investigated in (tapering) channels *on chip* [47] as well as in free jets (**Figure 3K-M**) [48].

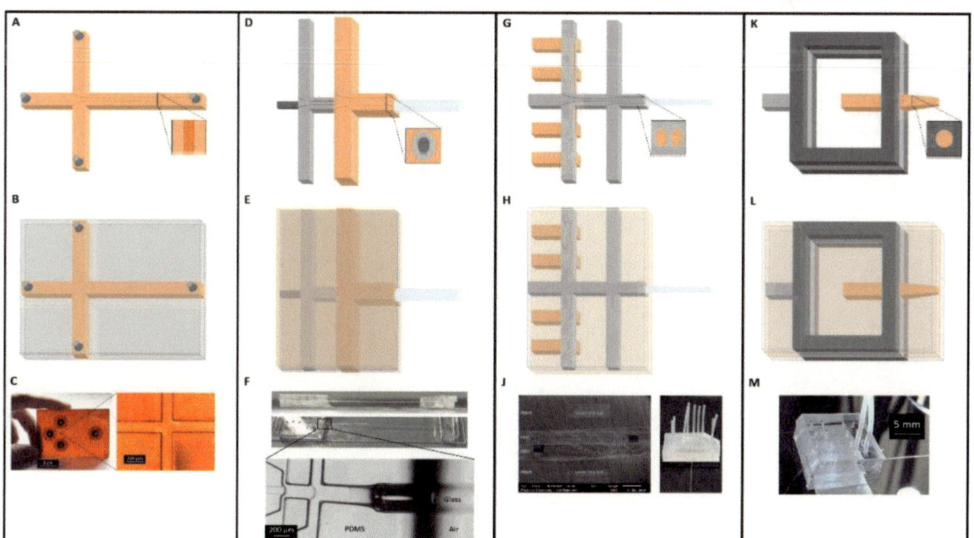

Figure 3.
Schematics and images of microfluidic devices used for time-resolved nanoparticle nucleation and growth and macromolecular self-assembly. (A) T-shaped, single layer (B) hydrodynamic focusing microfluidic device, made from Kapton (C) [40]. *(D) Three-dimensional (multilayer) hybrid hydrodynamic focusing device (E), made from PDMS with inserted glass capillary (F)* [24]. *(G) Double stream hydrodynamic focusing device, which can be aligned without optical access (H), made from SIFEL with a thin PDMS carrier layer and inserted glass capillary (J)* [23]. *(K) Multilayer (L) micro-jet device with hydrodynamic focused spray out of a nozzle, made entirely from PDMS (M)* [41].

4. Conclusion and outlook

Many hypotheses posed by researchers across the world would be answered by observing reactions in real time. To achieve this, two major technologies provide significant opportunity when merged: SAXS and optimised microfluidic setups. Although the last decades of research lead to significant developments in combining these fields, it remains a demanding intersection of methods with the potential to answer many fundamental questions around nucleation, growth and self-assembly of materials on the nanoscale. New three-dimensional hydrodynamic focusing device designs show great promise in studying a variety of systems – from organic to inorganic and crystal growth to self-assembly processes. Nevertheless, this is still an emerging field, with microfluidic and synchrotron technologies continuing to push the boundaries of possible experiments, opening up new possibilities for further reducing dead times, and thereby understanding the earliest parts of synthesis reactions, which are of critical importance in future control and modification of nanoparticles for a wide variety of purposes. Micro-focused X-ray beams and beamline optimisation, meanwhile, will be a key component to access faster time-scales with SAXS. We hope that this overview of microfluidics and SAXS analyses, along with some of our insights will aid future investigations into this challenging, but exciting field.

Acknowledgements

The authors greatly thank Stephen Mudie and Calum Kinnear for useful remarks and comments.

Conflict of interest

The chapter was written through contributions of all authors. All authors have given approval of the version of the chapter. The authors declare no competing financial interest.

Author details

Susanne Seibt* and Timothy Ryan
Australian Synchrotron, ANSTO, Melbourne, Victoria, Australia

*Address all correspondence to: seibts@ansto.gov.au

References

[1] Manz A, Graber N, Widmer HM. Miniaturized total chemical analysis systems: A novel concept for chemical sensing. Sensors and Actuators B: Chemical. 1990;1:244–8.

[2] Kopp MU, Mello AJ, Manz A. Chemical amplification: continuous-flow PCR on a chip. Science. 1998;280: 1046–8.

[3] Harrison DJ, Fluri K, Seiler K, Fan Z, Effenhauser CS, Manz A. Micromachining a miniaturized capillary electrophoresis-based chemical analysis system on a chip. Science. 1993; 261:895–7.

[4] Hadd AG, Raymond DE, Halliwell JW, Jacobson SC, Ramsey JM. Microchip device for performing enzyme assays. Anal Chem. 1997;69:3407–12.

[5] Woolley AT, Mathies RA. Ultra-high-speed DNA sequencing using capillary electrophoresis chips. Anal Chem. 1995; 67:3676–80.

[6] Manz A, Harrison DJ, Verpoorte EMJ, Fettinger JC, Paulus A, Lüdi H, et al. Planar chips technology for miniaturization and integration of separation techniques into monitoring systems. Journal of Chromatography A. 1992;593:253–8.

[7] Ingham B. X-ray scattering characterisation of nanoparticles. Crystallography Reviews. 2015;21:229–303.

[8] Squires TM, Quake SR. Microfluidics: Fluid physics at the nanoliter scale. Reviews of Modern Physics. 2005;77: 977–1026.

[9] Neto C, Evans DR, Bonaccurso E, Butt H-J, Craig VSJ. Boundary slip in Newtonian liquids: a review of experimental studies. Reports on Progress in Physics. 2005;68:2859–97.

[10] Lee CY, Chang CL, Wang YN, Fu LM. Microfluidic mixing: a review. Int J Mol Sci. 2011;12:3263–87.

[11] Ward K, Fan ZH. Mixing in microfluidic devices and enhancement methods. J Micromech Microeng. 2015;25.

[12] Capretto LC, W; Hill, M; Zhang,X. Micromixing Within Microfluidic Devices. Microfluidics. Berlin, Heidelberg: Springer; 2011. p. 27–68.

[13] Feigin LA, Svergun DI, Taylor GW. Structure Analysis by Small-Angle X-Ray and Neutron Scattering. Boston, MA: Springer US; 1987.

[14] Bruus H. Theoretical microfluidics. Oxford: Oxford University Press; 2008.

[15] Lu M, Ozcelik A, Grigsby CL, Zhao Y, Guo F, Leong KW, et al. Microfluidic Hydrodynamic Focusing for Synthesis of Nanomaterials. Nano Today. 2016;11:778–92.

[16] McDonald JC, Duffy DC, Anderson JR, Chiu DT, Wu H, Schueller OJA, et al. Fabrication of microfluidic systems in poly (dimethylsiloxane). Electrophoresis. 2000;21:27–40.

[17] Duffy DC, McDonald JC, Schueller OJ, Whitesides GM. Rapid Prototyping of Microfluidic Systems in Poly(dimethylsiloxane). Anal Chem. 1998;70:4974–84.

[18] Xia Y, Whitesides GM. Soft Lithography. Angewandte Chemie International Edition. 1998;37: 550–75.

[19] Heckele M, Bacher W, Müller KD. Hot embossing - The molding technique for plastic microstructures. Microsystem Technologies. 1998;4:122–4.

[20] Becker H, Heim U. Hot embossing as a method for the fabrication of polymer high aspect ratio structures. Sensors and Actuators A: Physical. 2000;83:130–5.

[21] Amin R, Knowlton S, Hart A, Yenilmez B, Ghaderinezhad F, Katebifar S, et al. 3D-printed microfluidic devices. Biofabrication. 2016;8:022001.

[22] Bishop GW. 3D Printed Microfluidic Devices. Microfluidics for Biologists: Springer; 2016. p. 103–13.

[23] Fuerst C. Kinetic studies of lyotropuc structure formation with microfluidics and small angle X-ray scattering [Dissertation]. Bayreuth: University of Bayreuth; 2016.

[24] Seibt S, Mulvaney P, Förster S. Millisecond CdS nanocrystal nucleation and growth studied by microfluidics with in situ spectroscopy. Colloids and Surfaces A: Physicochemical and Engineering Aspects. 2019;562:263–9.

[25] Seibt S. In-situ Investigations of Molecular Self-Assembly Using Microfluidics [Dissertation]. Bayreuth, Melbourne: University of Bayreuth, The University of Melbourne; 2018.

[26] Chen X, Schroder J, Hauschild S, Rosenfeldt S, Dulle M, Forster S. Simultaneous SAXS/WAXS/UV-Vis Study of the Nucleation and Growth of Nanoparticles: A Test of Classical Nucleation Theory. Langmuir. 2015;31: 11678–91.

[27] Herbst M, Hofmann E, Forster S. Nucleation and Growth Kinetics of ZnO Nanoparticles Studied by in Situ Microfluidic SAXS/WAXS/UV-Vis Experiments. Langmuir. 2019;35: 11702–9.

[28] Polte J, Ahner TT, Delissen F, Sokolov S, Emmerling F, Thunemann AF, et al. Mechanism of gold nanoparticle formation in the classical citrate synthesis method derived from coupled in situ XANES and SAXS evaluation. J Am Chem Soc. 2010;132:1296–301.

[29] Harada M, Tamura N, Takenaka M. Nucleation and Growth of Metal Nanoparticles during Photoreduction Using In Situ Time-Resolved SAXS Analysis. The Journal of Physical Chemistry C. 2011;115:14081–92.

[30] Henkel A, Schubert O, Plech A, Sönnichsen C. Growth Kinetic of a Rod-Shaped Metal Nanocrystal. The Journal of Physical Chemistry C. 2009;113: 10390–4.

[31] Garcia PRAF, Prymak O, Grasmik V, Pappert K, Wlysses W, Otubo L, et al. An in situ SAXS investigation of the formation of silver nanoparticles and bimetallic silver–gold nanoparticles in controlled wet-chemical reduction synthesis. Nanoscale Advances. 2020;2:225–38.

[32] Abecassis B, Testard F, Spalla O, Barboux P. Probing in situ the nucleation and growth of gold nanoparticles by small-angle X-ray scattering. Nano Lett. 2007;7:1723–7.

[33] Abecassis B, Testard F, Kong Q, Francois B, Spalla O. Influence of monomer feeding on a fast gold nanoparticles synthesis: time-resolved XANES and SAXS experiments. Langmuir. 2010;26:13847–54.

[34] Polte J. Fundamental growth principles of colloidal metal nanoparticles – a new perspective. CrystEngComm. 2015;17:6809–30.

[35] Polte J, Erler R, Thunemann AF, Sokolov S, Ahner TT, Rademann K, et al. Nucleation and growth of gold nanoparticles studied via in situ small angle X-ray scattering at millisecond time resolution. ACS Nano. 2010;4: 1076–82.

[36] Polte J, Erler R, Thunemann AF, Emmerling F, Kraehnert R. SAXS in combination with a free liquid jet for improved time-resolved in situ studies of the nucleation and growth of nanoparticles. Chem Commun (Camb). 2010;46:9209–11.

[37] Gerstner D, Kraus T. Rapid nanoparticle self-assembly at elevated temperatures. Nanoscale. 2018;10: 8009–13.

[38] Merkens S, Vakili M, Sanchez-Iglesias A, Litti L, Gao Y, Gwozdz PV, et al. Time-Resolved Analysis of the Structural Dynamics of Assembling Gold Nanoparticles. ACS Nano. 2019;13: 6596–604.

[39] Nette J, Howes PD, deMello AJ. Microfluidic Synthesis of Luminescent and Plasmonic Nanoparticles: Fast, Efficient, and Data-Rich. Advanced Materials Technologies. 2020;5.

[40] With S, Trebbin M, Bartz CB, Neuber C, Dulle M, Yu S, et al. Fast diffusion-limited lyotropic phase transitions studied in situ using continuous flow microfluidics/ microfocus-SAXS. Langmuir. 2014;30: 12494–502.

[41] Trebbin M, Kruger K, DePonte D, Roth SV, Chapman HN, Forster S. Microfluidic liquid jet system with compatibility for atmospheric and high-vacuum conditions. Lab Chip. 2014;14: 1733–45.

[42] Schiener A, Wlochowitz T, Gerth S, Unruh T, Rempel A, Amenitsch H, et al. Nucleation and growth of CdS nanoparticles observed by ultrafast SAXS. MRS Proceedings. 2013;1528.

[43] Stegelmeier C, Exner A, Hauschild S, Filiz V, Perlich J, Roth SV, et al. Evaporation-Induced Block Copolymer Self-Assembly into Membranes Studied by in Situ Synchrotron SAXS. Macromolecules. 2015;48:1524–30.

[44] Haynl C, Hofmann E, Pawar K, Forster S, Scheibel T. Microfluidics-Produced Collagen Fibers Show Extraordinary Mechanical Properties. Nano Lett. 2016;16:5917–22.

[45] Hofmann E, Kruger K, Haynl C, Scheibel T, Trebbin M, Forster S. Microfluidic nozzle device for ultrafine fiber solution blow spinning with precise diameter control. Lab Chip. 2018;18:2225–34.

[46] Dehsorkhi A, Castelletto V, Hamley IW, Adamcik J, Mezzenga R. The effect of pH on the self-assembly of a collagen derived peptide amphiphile. Soft Matter. 2013;9.

[47] Trebbin M, Steinhauser D, Perlich J, Buffet A, Roth SV, Zimmermann W, et al. Anisotropic particles align perpendicular to the flow direction in narrow microchannels. Proc Natl Acad Sci U S A. 2013;110:6706–11.

[48] Schlenk M, Hofmann E, Seibt S, Rosenfeldt S, Schrack L, Drechsler M, et al. Parallel and Perpendicular Alignment of Anisotropic Particles in Free Liquid Microjets and Emerging Microdroplets. Langmuir. 2018;34: 4843–51.

Chapter 3

Prototyping and Production of Polymeric Microfluidic Chip

Honggang Zhang, Haoyang Zhang, Tianyu Guan, Xiangyu Wang and Nan Zhang

Abstract

Microfluidic chips have found many advanced applications in the areas of life science, analytical chemistry, agro-food analysis, and environmental detection. This chapter focuses on investigating the commonly used manufacturing technologies and process chain for the prototyping and mass production of microfluidic chips. The rapid prototyping technologies comprising of PDMS casting, micro machining, and 3D-printing are firstly detailed with some important research findings. Scaling up the production process chain for microfluidic chips are discussed and summarized with the perspectives of tooling technology, replication, and bonding technologies, where the primary working mechanism, technical advantages and limitations of each process method are presented. Finally, conclusions and future perspectives are given. Overall, this chapter demonstrates how to select the processing materials and methods to meet practical requirements for microfluidic chip batch production. It can provide significant guidance for end-user of microfluidic chip applications.

Keywords: microfluidic chips, micro structure, mold insert, replication, microinjection molding, bonding

1. Introduction

On the basis of channels with dimensions of tens to hundreds of micrometers, microfluidics principally deals with the processing and manipulation of tiny amounts (normally from 10^{-9} to 10^{-18} liters) of fluids [1]. Although early microfluidic devices relied mainly on silicon and glass, the use of polymer materials has become increasingly common, which is largely attributed to its relatively low cost, admirable replication accuracy, optical transparency, biocompatibility, chemical stability, and good electrical insulation. Integrating with other apparatuses such as detectors and purifiers, polymeric microfluidics contains micro components, which principally includes microchannels [2, 3], microvalves [4–6], micropumps [7], micromixers [3, 8]. In fact, the elementary surface structures, which generally include channels, wells, chambers, and protruding arrays in microscale and submicro scale, are crucial for the functionalities of the entire microfluidic systems. The cross-sectional shape is normally square, and the aspect ratio is commonly less than two with a surface roughness being smaller than 50 nm.

For mass production of a microstructured plastic part for microfluidic applications, an appropriate tooling technology with the capability of enabling the fabrication of multi-scale features and controllable surface quality is required, because the feature size and quality of microinjection molded microfluidic chip is highly related to the characteristics and quality of corresponding micro mold insert [9–12]. The essence of replication is the reproduction of microfluidic structures from the master to the substrate materials. In terms of the replication of surface structures using polymer materials, techniques such as injection molding [13, 14] (including microinjection molding, variotherm-assisted injection molding, and injection compression molding), embossing [15, 16] (hot embossing, UV embossing, roll-to-roll embossing), nanoimprinting [17] and 3D printing [18] are commonly applied. Among all the replication processes, injection molding and hot embossing are more viable for industrial production, due to their advantages of relatively low cost and suitability for different materials and product design. Injection molding can be particularly efficient for mass production.

After microinjection molding of microfluidic devices, sealing is required to achieve enclosed microchannels [19]. Some important parameters should be taken into consideration before selecting bonding methods. Bond strength should be one of the most important parameters. In some applications, the interfacial bond energy is expected to be as high as the cohesive strength of the substrate, while in other applications the weak and reversible bonds between the cover and the substrate are required. Another parameter should be the solubility of thermoplastic and solvent. The interfaces used for bonding should be compatible with some solvent so that they can be dissolved and bonded together, meanwhile, the microchannels should not be subjected to the deformation in the bonding process [19]. Other parameters include the surface roughness, optical properties as well as material compatibility. In general, bonding methods can be either indirect or direct. In the indirect bonding process, an intermediate adhesive layer is used to bond two substrates together [20]. The interfaces applied with adhesive will have different properties than the bulk substrate. In terms of direct bonding, no other material is added between the interfaces, and the surface of the substrates can be mated directly [21]. After bonding, the interfaces and bulk material have homogenous properties.

This chapter will overview these prototyping and mass production technologies and process chains for manufacturing of polymer microfluidic chips.

2. Rapid prototyping of microfluidic chips

In recent years, microfluidic devices find many advanced applications in chemical analysis [22], polymerase chain reaction (PCR) [23], biological analysis [24], and chemical synthesis [25]. Main rapid prototyping methods of fabricating microfluidic chips include PDMS casting, micromachining, and 3D printing, etc.

2.1 PDMS casting

PDMS casting is also named soft lithography. The polydimethylsiloxane (PDMS) is a kind of silicon-based organic polymer material that has been widely used in microfluidic devices by rapid prototyping [26]. PDMS casting fabrication process is typically divided into the following steps (see **Figure 1**): molds developing, PDMS casting, curing and releasing, bonding, and integration [14]. A master mold needs to be prepared firstly, where SU-8 epoxy resin usually is applied as the mold material [27]. The standard PDMS compositions are composed of silicon elastomer base and the curing reagent in a ratio of 10:1. The uncured PDMS is poured into the mold, followed by curing at 70–80 °C for an hour. After releasing PDMS from the mold, PDMS microfluidic chips can be obtained [28]. The post-process of PDMS

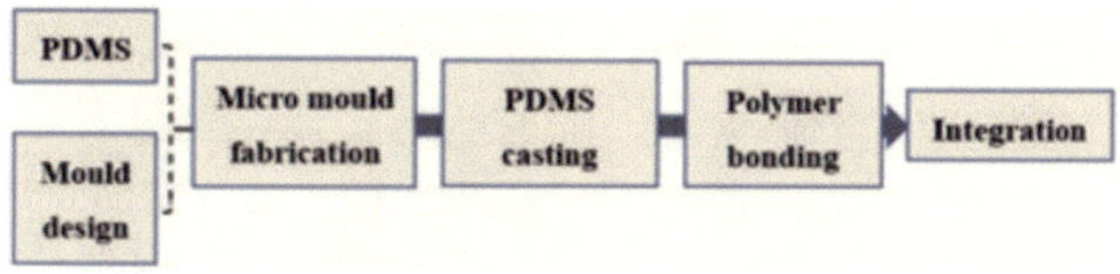

Figure 1.
Process steps of PDMS casting.

casting usually includes bonding and integration. Bonding can reduce the hydrophobicity of the PDMS chip to encapsulate its microchannel. To enhance the bonding strength of chips with other materials, adjusting bonding process parameters and surface modification by using oxygen plasma to form O-Si-O covalent bonds at the interface of PDMS microfluidic channels are commonly used [29]. Finally, some microsensors, microheaters and microfluidic pumps are integrated onto microfluidic chip for diversified performance [18]. **Figure 2** shows the practical microfluidic chip fabrication approaches by the PDMS casting process.

Although PDMS casting is a rapid prototyping process for disposable microfluidic chip fabrication, it is a complex process with many drawbacks. The microfluidic chip fabricated by PDMS casting has insufficient mechanical strength, non-conductivity, and non-magneticity. Also, under high temperature, high voltage, and pressure conditions, it is not easy to integrate other precision components on the microfluidic chip [19].

2.2 Micro machining

The microfluidic chip fabricated by micro machining, such as femtosecond laser or endmill, has relatively high accuracy and can be used repeatedly [20]. Combing with computer numerical control (CNC), micro machining can rapidly and accurately construct devices at several microns scale [21]. However, microchannels obtained from micro milling are too rough to use as a microfluidic chip, especially for the machining of polymer materials [22]. However, compared with micro milling, laser micro machining with improved accuracy can perform the direct writing ablation of polymers such as PMMA, PC, etc. Direct-write laser machining with UV and CO_2 assistance can realize microstructuring in polymer materials.

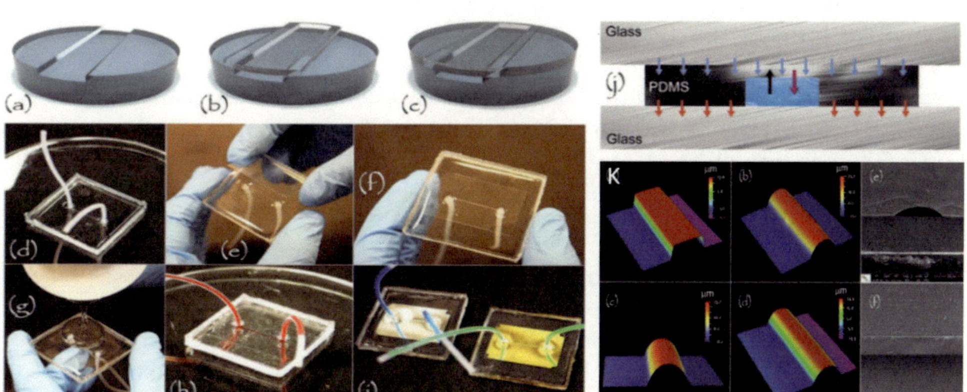

Figure 2.
PDMS casted microfluidic chip: PDMS substrate with microchannels (a); enclosing of microchannel (b, e, f); PDMS molding(g); PDMS curing (c, h); substrate integration to fluid pump (d); SWB was applied for prototyping PDMS (I; forces in reversible bonded microdevice (j), mold characterization and cross-section (k) [30].

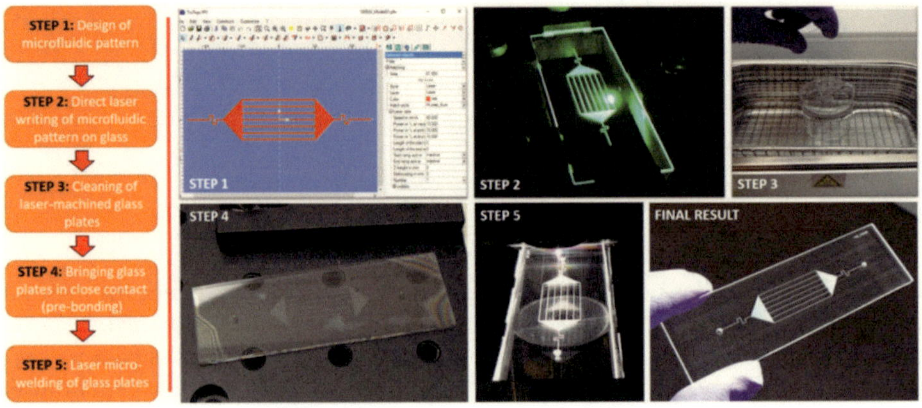

Figure 3.
Direct-write laser machining for microfluidic chip [31].

For laser micro machining, when the laser beam is moving above the workpiece space, the heated spots will gather to form various patterns. Through the reflection in the light path and the focusing effect of the focusing lens, the laser is finally focused on the workpiece, where the temperature of the focal point increases rapidly. After the polymer material melts and decomposes, scratches will be left on the surface of the polymer, where the microchannels are formed [24]. The scanning speed is programmed and controlled by a computer. **Figure 3** shows the specific process methods of laser micro machining and machined glass microfluidic chip.

2.3 3D-printing

3D-printing is different from traditional manufacturing techniques. It achieves the fabrication of materials utilizing additive manufacturing (AM). Under computer-aided control, it can construct the 3D structure layer by layer. The most common 3D printing technologies used in the manufacture of microfluidic devices are stereo-lithography (SL), multi-jet modeling (MJM), and fused deposition modeling (FDM). SL utilizes selective light exposure to photopolymerize precursor to construct object layer by layer [27]. For MJM, it works by using an inkjet head to spray curable liquid photopolymers into a tray, and photopolymerization will happen on each layer when exposed quickly to UV light. FDM uses a motor-driven nozzle head to print heated thermoplastic material in three dimensions. **Figure 4** demonstrates the 3d-printing devices and printed polymer microfluidic chips. A thin resin layer as printed material is solidified via laser beam for the fabrication of 3D-chip that features the channel layer and bottom layer of 500 μm and intersects at 45° and 20°. The printing resolution is 50 μm in line width. After the printing process, the chip needs to be washed using isopropanol (IPA) and deionized (DI) water in the micropump platform.

3. Scaling up production of microfluidic chips

In general, to achieve batch production of polymer microfluidic chips, a high-quality mold insert is indispensable for precision replication of micro structure using microinjection molding. In this case, developing a reliable tooling technology for mold insert fabrication is important. Besides, after microinjection molding of polymeric chips, the bonding process for sealing microchannels of chips determines the final functionality of microfluidic chips. Therefore, each process step is significantly important.

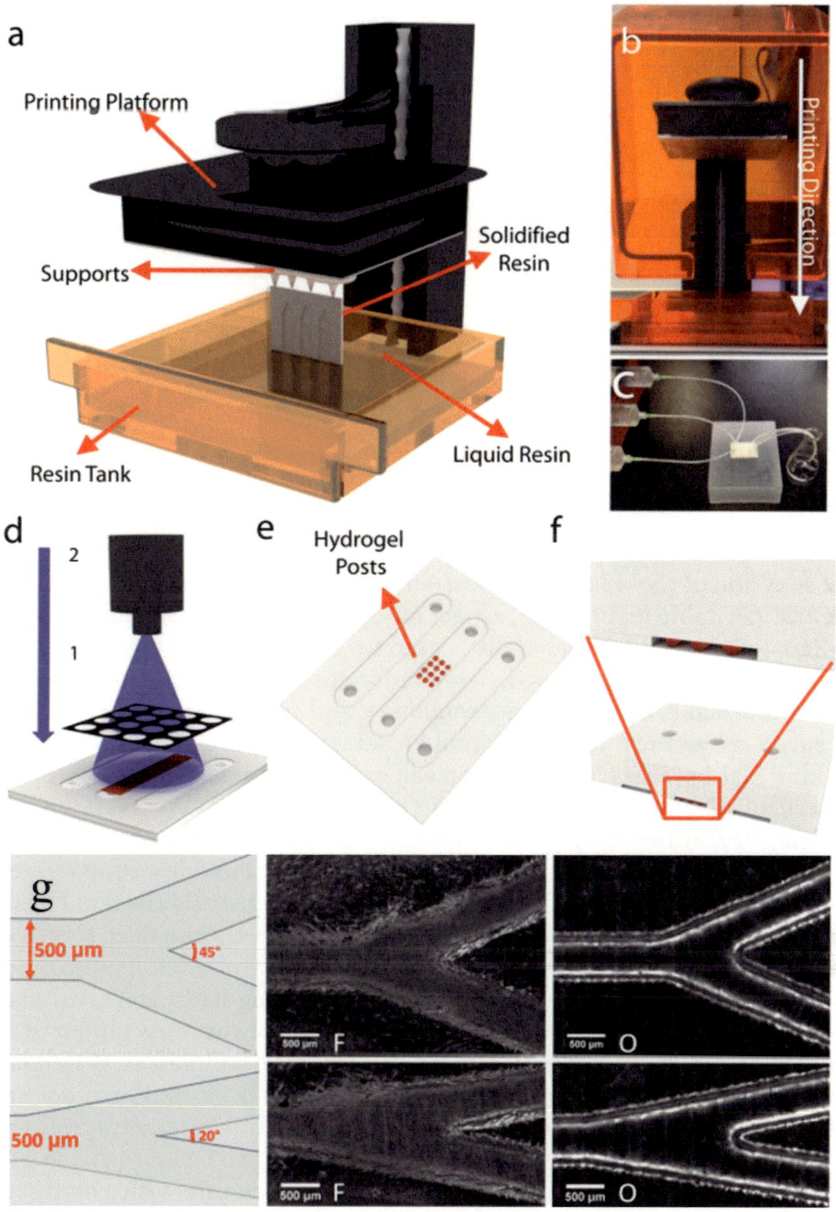

Figure 4.
3D printed polymer microfluidic chip: (a)schematic of SLA process; (b) printing platform configuration; (c) channel washing; (d) application of hydrogel; (e) remaining hydrogel remove; (f) cross-section of the 3D chip; (g) 500 µm raised structures intersecting at a 45° and 45° [32]. Copyright (2016) with permission from IOP publishing, LTD.

3.1 Tooling technology

There are a variety of manufacturing technologies available to fabricate micro mold inserts, such as ultraprecision micro milling, micro-electrical discharge machining (µEDM), electrochemical machining (ECM), silicon wet etching, deep reactive ion etching, laser machining, and LIGA-based processes (LIGA, UV-LIGA). **Table 1** shows a comparison among the mold insert manufacturing technologies according to the achievable feature size, surface roughness, aspect ratio, and machinable materials.

Technology	Feature size (μm)	Surface roughness (μm)	Aspect ratio	Material
Micro milling	25 ~ 100	0.2 ~ 5	10	Brass, COC, steel
μEDM	10 ~ 25	0.05 ~ 1	50 ~ 100	Conductive materials
ECM	10	0.02	NA	Conductive materials
Laser machining	1 ~ 5	0.4 ~ 1	<50	Any
X-ray lithography	0.5	0.02	100	Photoresist
UV lithography	0.7 ~ 1.5	NA	22	Photoresist
Deep reactive ion etching	2	NA	10 ~ 20	Silicon
Electroforming	0.3	<10	<10	Copper/nickel/alloy

Table 1.
Comparison between various manufacturing techniques for the fabrication of microstructured mold insert [33].

The selection of mold insert fabrication technology is based on the geometrical complexity, desirable feature size, aspect ratio, surface roughness, and processing cost. Micro milling or EDM generally is used for fabricating micro structure with feature sizes larger than 50 μm with a tolerance of several micrometers. However, it is difficult to achieve a low surface roughness and some sharp corners processing. **Figure** 5 shows the microfluidic mold insert fabricated by micro milling and die-sinking EDM process. LIGA, silicon wet etching, and deep reactive ion etching have excellent advantages for sub-micron fabrication [34]. ECM is suitable for structuring 3D features with less sub-surface damage and higher dimensional precision at the nanometric range [35]. The selection of mold insert materials is also critical. When the production yield is less important and the mold insert lifetime is not critical, silicon wafer mold insert fabricated from deep reactive ion etching is favorable to the precision replication of plastic microchip with optical surface finishing [36]. However, in most instances, a long-life metal mold insert having high wear resistance and hardness is desirable for the mass production of microinjection molded microfluidic chips. For example, stainless steel mold insert can be against the feature being worn or surface quality degradation under up to tens of thousands of microinjection molding cycles. In this case, ultraprecision machining, LIGA process, laser machining, and μEDM can be more viable. Also, the LIGA process and μEDM can fabricate the micro structure with a high aspect

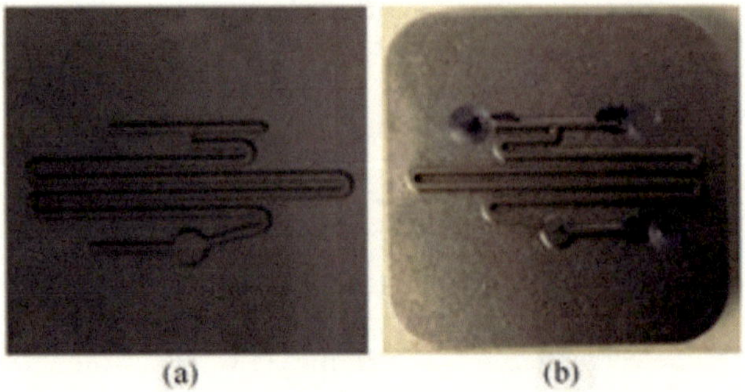

Figure 5.
Micro milling and die-sinking EDM fabricated mold insert: graphite electrode (a) and stainless tools (b) [33]. Copyright (2015) with permission from IOP publishing, LTD.

ratio and tight tolerance [37, 38]. **Figure 6** shows the dry-etched silicon wafer mold, UV lithographic photoresist mold, and electroformed nickel mold, which were fabricated by the author' team. In the following sections, the main manufacturing technologies for mold insert fabrication are detailed.

3.1.1 LIGA

LIGA technique comprising of lithography, electroforming, and molding is a multi-step replication process to generate micro structure with the desired patterns [39]. It has been a promising technology for industrial-scale commercialization [40, 41]. The typical process methods follow the consequential steps below: 1) the photoresist (AZ or epoxy resin SU-8) is firstly evenly coated on the silicon wafer substrate along with the subsequent baking process; 2) a pre-prepared photomask with the desired patterns is placed on the top surface of the photoresist at a good alignment manner for further irradiation exposure; 3) the exposed areas is removed/remained chemically using developer (depending on the type of photoresist), where the patterns are transferred to silicon wafer from mask and the dimensional accuracy of patterns can be controlled by lithography parameters (exposure time and exposure dose); 4) seed layers of adhesive layer (Ti/Cr) and conductive layer (Au/Ni) are sputtered onto the structured photoresist surface for metallization; 5) the following step is electroforming for fabricating a microstructured mold insert, where the metallized patterns on silicon wafer serve as a cathode for nickel deposition; after electroforming, a electroformed replica is relived via silicon chemically etching and photoresist is chemically removed; the final replica can be used as a mold insert; 6) such an electroformed mold insert can be used as a master for replication of microfluidic chips by microinjection molding process [42]. **Figure 7** details the specific process steps of mold insert fabrication in various microstructuring technologies assisted by electroforming.

Due to costly x-ray synchrotron for X-ray lithography, the LIGA process is not commonly used in the industrial field. As an alternative, UV lithography has been a favorable technology to prepare master for electroforming [37, 43]. As a result, UV-LIGA has been an acceptable process to fabricate the mold insert with the

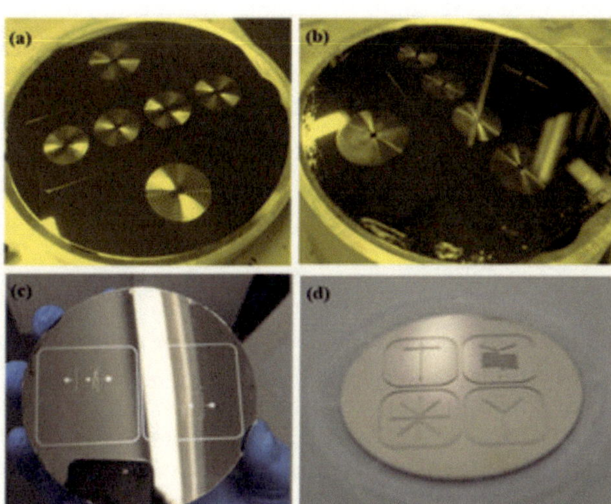

Figure 6.
(a) Dry etched silicon wafer; (b) UV lithographic photoresist master; (c) and (d) electroformed nickel mold insert. Copyright (2020) with permission from IOP publishing, LTD.

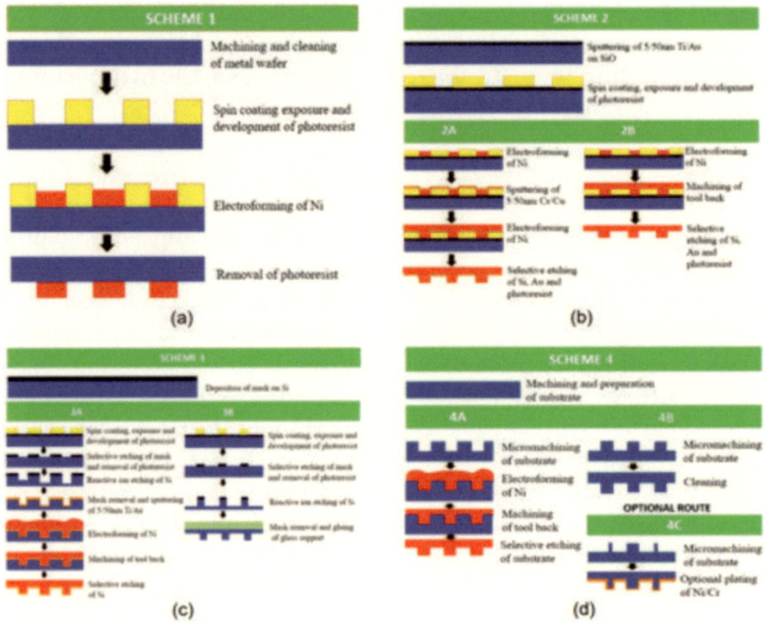

Figure 7.
Basic schemes of manufacturing of mold insert.: (a) metal substrate, lithography, and electroforming; (b) a silicon substrate, lithography, and electroforming; (c) deep reactive ion etching of silicon (d) micro-machining of non-silicon substrates [9]. Copyright (2020) with permission from IOP publishing, LTD.

feature size of several micrometers to hundreds of micrometers, although the reachable aspect ratio of this technique is limited in the range of ~20 [34]. Considering the specific microfluidic applications, UV-LIGA is sufficient to fabricate the desired micro structures. Additionally, some LIGA-like processes, such as EUV-LIGA, EBL-LIGA, IB-LIGA, are also developed towards nanoscale micro structuring [44]. However, these LIGA-like techniques still have some shortcomings for mold insert fabrication. First, the mold materials usually are nickel and nickel alloy, the tool steel is not included due to the restriction from the electroforming process. Besides, the integration of draft angle onto mold insert is not easy, but it is critical for demoulding of the polymeric part from mold insert to reduce the potential demoulding deformation and damages of micro structures [45]. The research shows that the draft angle can be achieved by properly adjusting the UV/X-rays exposure dose.

3.1.2 Laser machining

The fabrication of a metallic mold insert is generally time consuming and expensive. Laser machining seems to be more competitive than other mechanical machining techniques. It allows the rapid fabrication of micro structure with a feature size of several micrometers at an aspect ratio of ~10 [46]. However, due to the limitation of spot size, making high precision micro mold insert with tight tolerance by direct laser micromachining is challenging. The minimum spot size is usually half wavelength of the light utilized [47]. In this context, the laser LIGA technology combining laser ablation and electroforming process is developed to produce precision mold insert with 3D structures, where laser ablation is applied to form polymer stencil for subsequent metallic replication by electroforming to fabricate nickel micro mold insert. **Figure 8** shows the microstructure metal mold insert and polymer mold insert fabricated by laser machining and laser ablation, respectively.

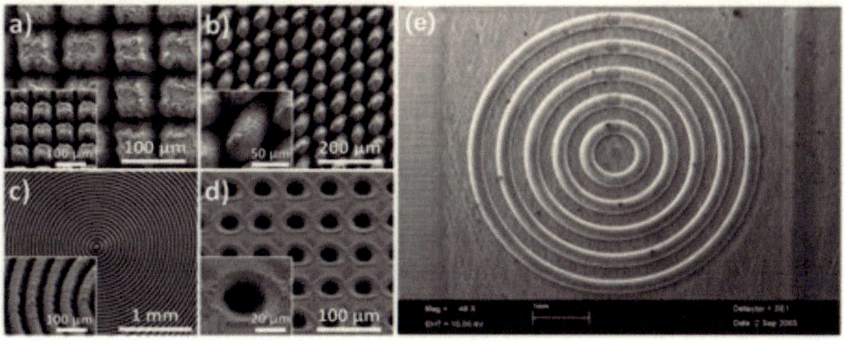

Figure 8.
(a-d) Micropatterns on metal mold insert fabricated by laser machining [48]; (e) Polymer mold insert fabricated by laser ablation [49]. Copyright (2004) with permission from Royal Society of Chemistry.

Of course, when micro mold insert with 3D structures is used for microinjection molding of the microfluidic chip, the use of conventional methods often brings challenges to microstructure demoulding.

3.1.3 Micro electrical discharge machining (μEDM)

The μEDM technology is a non-conventional electrochemical processing method. The mold insert is fabricated based on the material removal mechanism with the erosive effect of the electrical discharges that happen between electrode and workpiece. The electrode and workpiece are placed with a small enough gap so that the voltage–current system can efficiently ionize the dielectric. The shape and dimension of the processed workpiece are determined by the electrode pattern. Micro mold insert with the feature size of ~20 μm can be machined by μEDM technology but the forming accuracy and surface roughness are not able to be guaranteed. **Figure 9** demonstrates

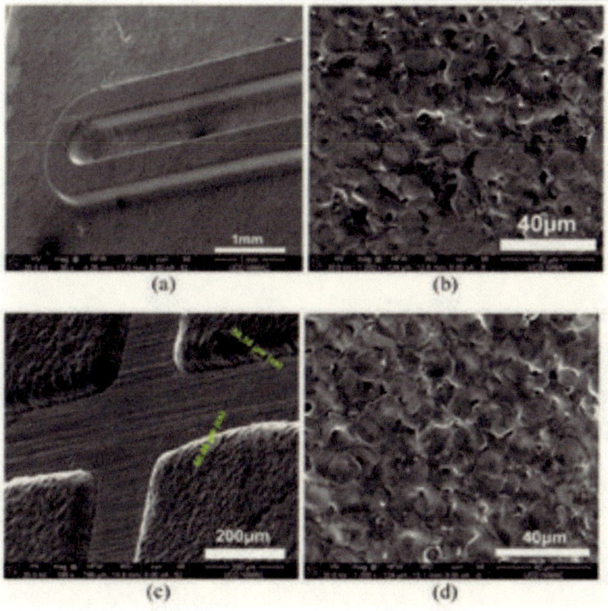

Figure 9.
Stainless steel mold insert fabricated by μEDM technology: graphite electrode with better geometry definition on micro-patterns (a); copper electrode with non-uniform edge finish (c); the crater size of both electrodes ((c) and (d)) [33]. Copyright (2015) with permission from IOP publishing, LTD.

the stainless steel mold insert machined by μEDM. With the assistance of other technologies, the surface finishing issues can be solved, such as ultrasonic-assisted μEDM. Overall, μEDM is a competitive technology, which has non-contact mechanical force, heat, and stress generation for machining the mold insert for microfluidic applications.

3.2 Micro injection molding of microfluidic structures

Micro injection molding is an important process to transfer patterns from mold to the surface of microfluidic patterns. Thanks to fast solidification and relatively high viscosity, filling of polymer melts into tiny micro patterned mold is challenging. Upon micro pattern is formed, separation of patterns from mold is problematic due to the distortion or damage of plastic micro structure from friction and adhesion between polymers. Here a brief introduction of these issues is presented by using a microfluidic chip as an example and a simulation strategy for microfluidic structure optimization is introduced.

3.2.1 Structures filling mechanism for micro injection molding

Figure 10(a) illustrates a fast prototyping mold for the injection molding of a flow cytometer chip, which has an area of 26.12 mm × 26.12 mm and a thickness of 1.05 mm [50]. **Figure 10(b)** shows surface structures that are used to fabricate inverted channels, all of which are 250 μm wide and 150 μm deep for the parts that had rectangular cross-sectional areas. The horizontal channel works as a fluidic channel, in which biological particles, such as cells, are focused by sheath flow from side inlets. Meanwhile, an optical fiber is integrated into the microfluidic chip using tilted channels. One channel is used for a fiber laser excitation beam and the other two are allocated for detection of forwarding scattering and side/fluorescence scattering by fibers. In order to minimize light intensity reduction of the excitation

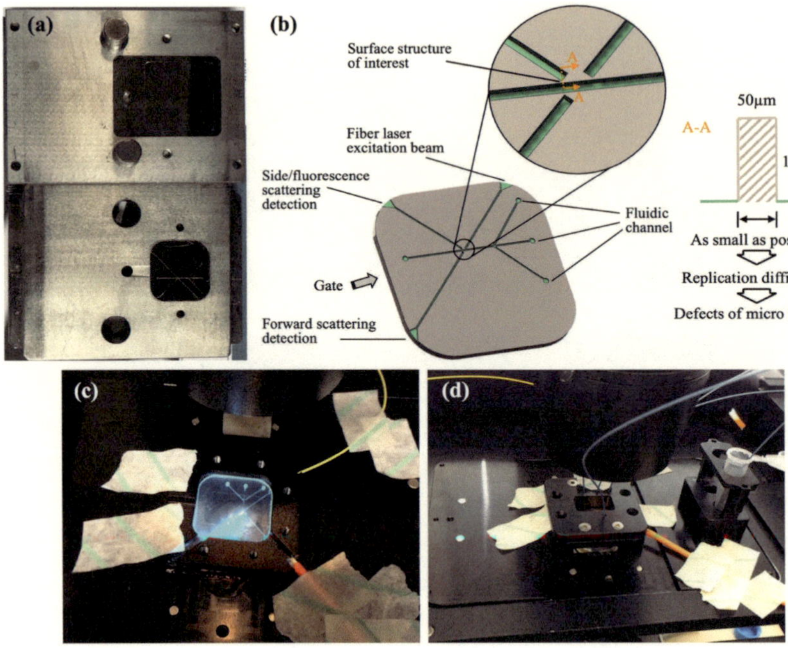

Figure 10.
Flow cytometer chip: (a) cassette mold, (b) three-dimensional model of surface structures on the chip, (c) actual chip assembly, (d) connection with tubing [55].

laser and to decrease potential scattering when the laser beam passes through plastic material, the distance between the fluidic channel and optical fiber channels has to be as small as possible. This requirement necessitates surface structures as shown in the upper image in **Figure 10(b)**. The protruding structures demonstrate a gradual aspect ratio, and the highest ones are at the places between the edge and the fluid channel, which have an aspect ratio of 3:1 nominally; it is these that were mainly investigated. **Figure 10(c)** shows the actual microfluidic chip (after bonding), which is assembled into a designed chip holder and connected with samples and PBS (Phosphate-buffered saline) sheath flow using PTFE (Polytetrafluoroethylene) tubing (**Figure 10(d)**).

In theory, the surface structures should fill more easily under such a layout because the gate was parallel to the feature of interest and the centre fluid channel [51]. However, replication of micro/nanoscale surface structures is challenging due to their high surface to volume ratios, especially for high aspect ratio features. Fast heat transfer rates mean that the polymer melts are inclined to solidify before the cavities are fully filled. For example, experimental results show that the cross-sectional area of the surface structure emphasized in **Figure 10(b)** only reaches up to 70.67% of the criterion. Combined with microscopy, process monitoring, and morphology, A combined melt flow and creep deformation model is proposed to explain the complex filling behavior of the surface structure. As shown in **Figure 11**, the overall replication is ordinarily composed of two parts: melt flow during the injection stage and creep deformation during the packing stage. Melt flow is related to injection velocity and pressure, while the extent of creep deformation is linked to mold temperature, packing pressure, packing time, etc. Based on this, increasing shot size to improve the replication quality of surface structures is proposed, and the replication is significantly improved and sufficient to satisfy the practical requirements of a microfluidic flow cytometer chip.

3.2.2 Process simulation and validation

In order to better understand the connection between product quality and process parameters, process simulation of injection molding has developed over several decades. Simulation in the early stages of part and mold design is relatively cost-efficient and offers the capability to evaluate various design options, such as runner design and gate designs. However, microscale effects, such as altering heat transfer coefficient (HTC), wall slip behavior, mold surface roughness, venting operations, which tend to be ignored in conventional injection molding simulation, should be considered in the simulation of microinjection molding effectively [52–54]. Another important concern is that most studies only used nominal machine processes, but they do not include the actual machine dynamics, which is important for microinjection molding of surface structures because microscale surface structures are much sensitive to process variation. As a result, simulation results are more or less unconvincing and cannot be adapted for real-world applications, and the simulation inaccuracy needs to be addressed.

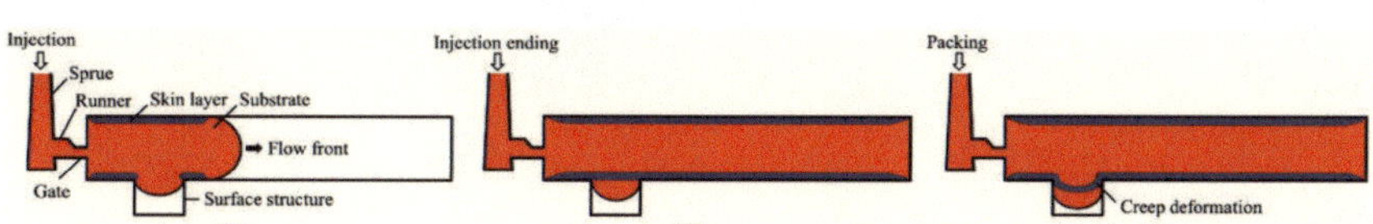

Figure 11.
Filling mechanism of surface structure on a substrate: (a) the injection stage, (b) the end of the injection stage, (c) the packing stage [50]. Copyright (2018) with permission from IOP publishing, LTD.

Based on an experimental study on the filling behavior of microfluidic surface structures (**Figures 10(b)** and **11(a)**), the real responses of the injection molding machine are acquired and adopted in the process settings of the simulation with the help of process monitoring [55]. In addition, the effect of microscale sensitive parameters on the replication of surface structures using simulation is systematically studied and validated by flow front profile, cross-sectional profile, and replication of the structures. Consequently, the combination including a relatively higher heat transfer coefficient (30,000 W/(m2·K)) of the injection stage, standard atmospheric pressure (0.1 MPa) as the initial air pressure of venting, 0.7 as the friction coefficient for wall slip and a freezing temperature of 20 degrees above the glass transition temperature is selected. In terms of the flow cytometer surface structures, replication defects in experiments (circled in **Figure 12(a)**) are successfully predicted after the optimization as the blue parts shown in **Figure 12(c)**. Besides, the insufficient replication of the droplet cylinders (the areas in white in **Figure 12(c)**) is also predicted after the selected parameters are applied.

3.3 Bonding techniques for microfluidic devices

The bonding methods for microfluidics comprises adhesive bonding, thermal fusion bonding, UV assisted thermal bonding, and solvent bonding. Following each section is introduced based on recent research findings with the mechanism, applicable materials, and process parameters of each bonding method. Finally, the challenges and relevant solutions for different bonding techniques are given.

3.3.1 Adhesive bonding

Adhesive bonding is a simple manner to seal microfluidic devices, compared to other bonding techniques. By applying the liquid adhesive onto the surface of the chips, the solvent composition starts to evaporate till two parts are bonded. Besides, some adhesives with epoxy or acrylate compositions need to be cured under UV light irradiation or upon heating. After mixing with the photo-initiator or catalyzing agent, polymerization or crosslink reaction can occur in the adhesive system [56], where the adhesive is cured and microfluidic devices are bonded.

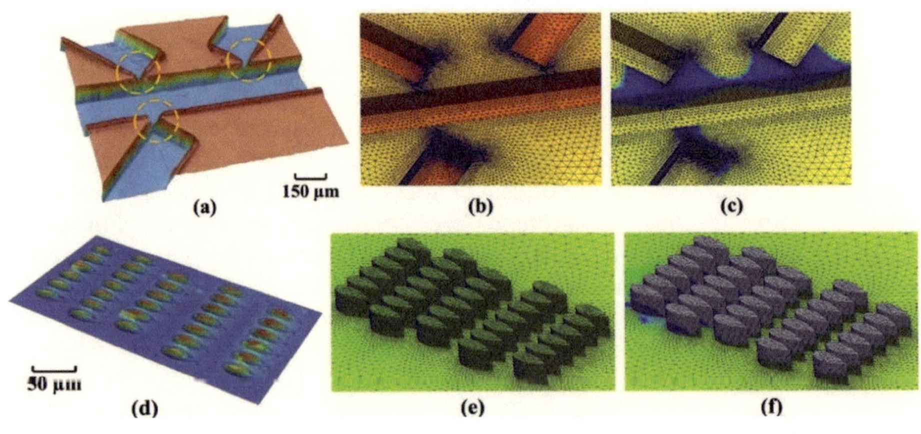

Figure 12.
Optimization and validation: (a) experimental result, (b) simulation result with default settings, and (c) simulation results (after the cycle including the packing and cooling stage) with optimized parameters for the replication of flow cytometer structures; (d) experimental result, (e) simulation result with default settings, and (f) simulation results (after the injection stage) with optimized parameters for the replication of microfluidic droplet cylinders [55].

UV-curable adhesive as an intermediate layer has also been developed to assemble the disposable PMMA chips [57]. The substrate is firstly washed by ultrasonic cleaning equipment with deionized (DI) water and flushed with nitrogen before bonding. Then the adhesive is applied to the cover plate by spin coating (**Figure 13b**). After spin coating, the cover plate is assembled with the substrate, then silicone tubes are connected to the inlet and outlet ports of the chips. According to **Figure 13c**, the isopropanol is injected into the inlet ports and comes out from the outlet port through the silicone tubes in the flushing process. This process is aimed at flushing out the uncured adhesive that has been trapped inside the microchannels to prevent the channel clogging. In the last stage, the assembled chips are exposed to the UV light irradiation and the polymer chains in the adhesive system are crosslinked. **Figure 14** shows the bonding state of a microfluidic chip with time using adhesive bonding technology.

Although this method is useful for disposable devices, the storage time of adhesives at room temperature should be highlighted. Meanwhile, using isopropanol to flush out the uncured adhesive is a complicated procedure, as it might also wash out the adhesive at the edge of the microchannels, then the bonding effect will be reduced. In this way, the burst pressure should be conducted in case the edge of the channels is not fully bonded.

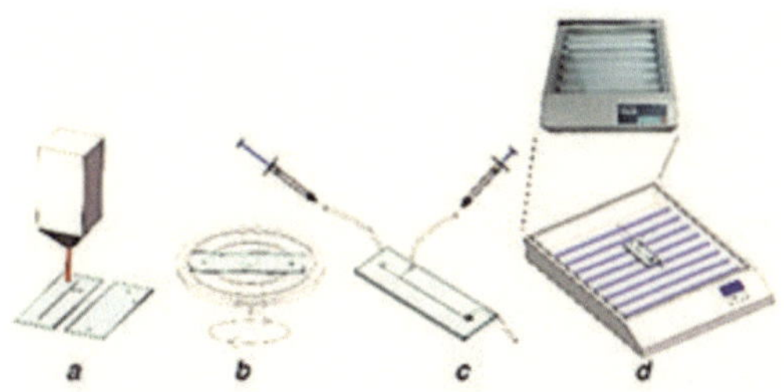

Figure 13.
Schematic diagram of the microchannel fabrication and bonding process: laser ablation for the fabrication of through holes on PMMA (a); spin coating of UV curable adhesive on cover plate (b); flushing out the un-crosslinked adhesives (c); UV exposure to induce the crosslinking of adhesives in the whole chip system [57].

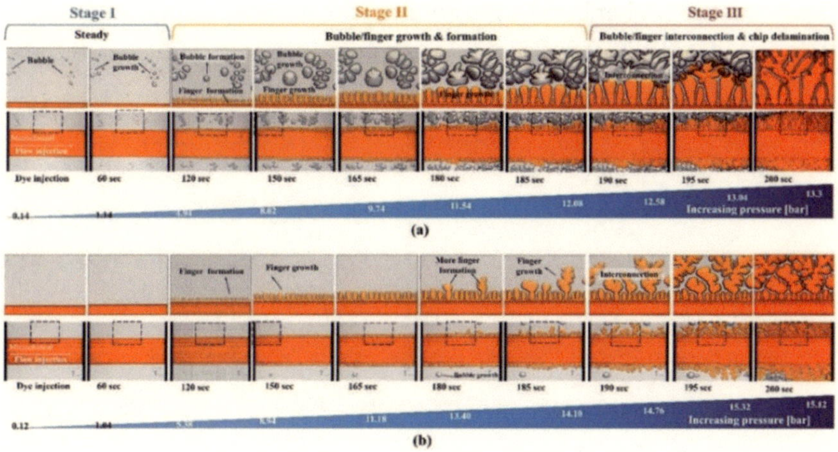

Figure 14.
The microfluidic chip bonding state with times [58].

3.3.2 Thermal fusion bonding

During the thermal fusion bonding process, the heating temperature is raised above the glass transition temperature (Tg) of the cover plate [19]. Meanwhile, a hold pressure is applied to enhance the mating contact forces between two surfaces. To achieve a strong bond, the heating temperature and the hold pressure should be high enough to ensure the complete diffusion of the polymer chains. In this case, the bond strength in the interface can be as high as the cohesive strength in the bulk material. According to **Figure 15**, the cover is first treated by O_2 plasma to make its surface more hydrophilic. In this case, the adhesion between the COC cover plate and PMMA substrate is greatly improved. During the thermal press process, heat and constant pressure are provided. Steel platen transfers both heat and pressure to the chips. Two rubber sheets are used to distribute the pressure evenly and two polyimide films are used for the anti-sticking purpose (**Figure 15**).

One critical problem of thermal bonding is that the temperature and the pressure cannot be too high, otherwise, the microchannels may collapse and their integrity cannot be well maintained. Therefore, the heating temperature, hold pressure, and hold time should be controlled and adjusted to achieve high bond strength and limit the channel deformation. The balance should be achieved among the heating temperature, hold time, and hold pressure. These parameters should be well adjusted to maintain the integrity of the channels.

3.3.3 UV assisted thermal bonding

The channel collapse is hard to avoid during thermal bonding, as the heating temperature is often higher than the Tg of material. Therefore, it is important to lower the heating temperature. In this case, the substrate or cover plate can be pre-treated by UV/Ozone to achieve bonding at low temperature. The pre-treated surface has higher energy due to oxidation. Therefore, the hydrophilicity and wettability are improved. As a result, the adhesion between the two surfaces is promoted.

Tsao et al. [60] compared the cross-section of the PMMA microchannels with and without UV/Ozone pre-treatment. As shown in **Figure 16(a)**, this chip treated by UV/Ozone had good bonding integrity. However, the untreated chips appeared obvious channel collapse as shown in **Figure 16(b)**. It is considered is because the treated chip had higher bond strength than the untreated chip (0.624 mJ/cm^2 compared to 0.003 mJ/cm^2).

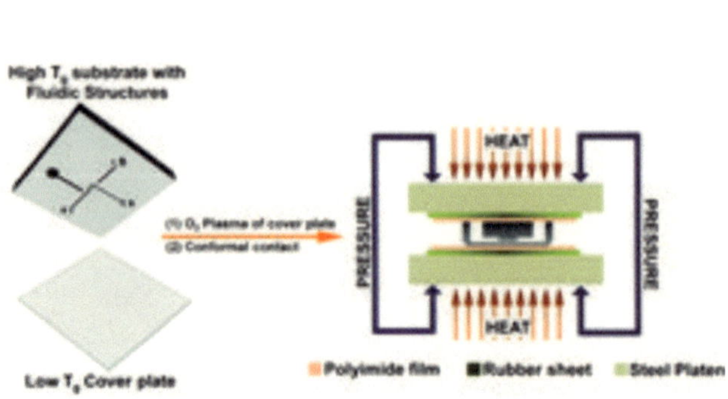

Figure 15.
Schematic of the protocol used for assembly of the hybrid-based fluidic devices and the thermal press instrument [59]. Copyright (2015) with permission from Royal Society of Chemistry.

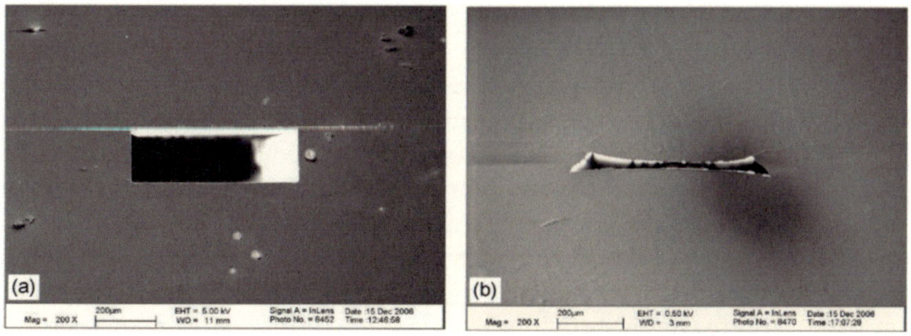

Figure 16.
SEM images of 500 mm wide, 180 mm deep PMMA microchannels: (a) thermal bonding of 24 min at 60°C, and (b) thermal bonding at 100°C [60]. Copyright (2007) with permission from Royal Society of Chemistry.

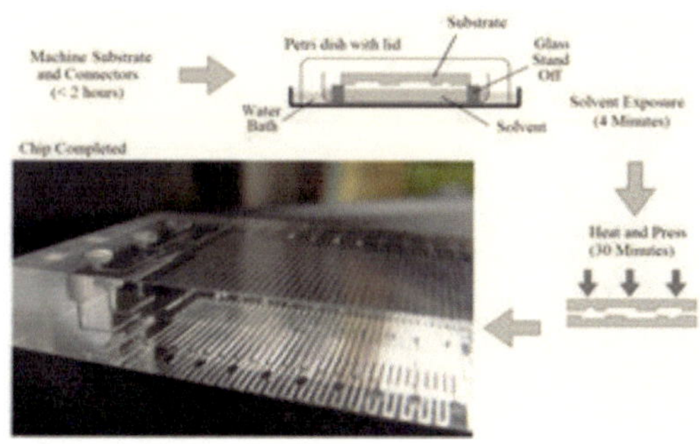

Figure 17.
Schematic of the bonding process. The PMMA chip shown has overall dimensions of 40[w] x 80[l] x 9.5 mm[h] [61].

3.3.4 Solvent bonding

Organic solvents can interact with polymer materials due to their similar solubility. Two polymer substrates are bonded together, which is called solvent bonding. This solubility can be determined by the Hildebrandt parameter (δ).

Ogilvie et al. [61] used chloroform vapor to bond the PMMA substrates. Two PMMA substrates were first exposed to chloroform vapor for 4 minutes. In this stage, the solvent was only 2–3 mm below the surface of the chip and both chips and solvent were covered by a glass lid to form a vapor exposure chamber (**Figure 17**). Then two substrates were bonded at 65 °C for 20 minutes, followed by the cooling process to room temperature in 10 minutes. All chips needed 12 hours of post-conditioning before use. The optical properties were well maintained after bonding and high bond strength was achieved. Solvent vapor is more controllable than traditional vapor bonding techniques such as dipping or soaking in the solvent. However, controlling the solvent concentration is necessary especially for solvent mixtures, as it will change during bonding.

4. Conclusions

In this chapter, the related manufacturing technologies for microfluidic chip fabrication are detailly described. Although rapid prototyping technologies for microfluidic chips, such as PDMS casting, micro machining, and 3D-printing, are well used in laboratory, the efficiency, machining accuracy and surface integrity of chips are still problematic for low-cost industrial batch production. Mass production based on micro injection molding is important for the fabrication of plastic microfluidic chips, where multiple metal micro mold tool inserts can be used. The LIGA-like process with the characteristic of high precision replication, such as UV-LIGA, presents considerable advantages to fabricate industrial-grade mold tool inserts for the fabrication of plastic microfluidic chips. For future development direction of microfluidic chip fabrication, a hybrid tooling technology for multi-scale mold insert should be explored to combine feature from micrometer scale to nanometer scale; high aspect ratio microchannel replication is an important task, which is challenging micro injection molding; the rheological behaviors of polymer materials are worth studying in nanoscale for high-precision microinjection molding of polymeric parts. Functional precision components, such as microsensors, micropumps, and microelectrodes, should be integrated onto microfluidic chips along with the bonding process. Additionally, quality control of microfluidic chips should be highlighted, such as channel dimensions and consistency.

Acknowledgements

The authors gratefully acknowledge the support from the Science Foundation Ireland (SFI) (No. 15/RP/B3208), the National Science Foundation of China (61675149), and the '111'project by the State Administration of Foreign Experts Affairs and the Ministry of Education of China (No. B07014).

Conflict of interest

The authors declare no conflict of interest.

Notes/thanks/other declarations

Thanks Honggang Zhang, Haoyang Zhang, Tianyu Guan, and Xiangyu Wang for their original draft writing and Dr. Nan Zhang for his review & editing, and supervision.

Author details

Honggang Zhang[1], Haoyang Zhang[1,2], Tianyu Guan[1], Xiangyu Wang[1] and Nan Zhang[1*]

1 Centre of Micro/Nano Manufacturing Technology (MNMT-Dublin), School of Mechanical and Materials Engineering, University College Dublin, Dublin, Ireland

2 State Key Laboratory of Precision Measuring Technology and Instruments, Centre of Micro/Nano Manufacturing Technology (MNMT), Tianjin University, Tianjin, China

*Address all correspondence to: nan.zhang@ucd.ie

References

[1] Whitesides, G.M., The origins and the future of microfluidics. Nature, 2006. 442(7101): p. 368. DOI: 10.1038/nature05058

[2] Anna, S.L. and H.C. Mayer, Microscale tipstreaming in a microfluidic flow focusing device. Physics of Fluids, 2006. 18(12): p. 121512. DOI: 10.1063/1.2397023

[3] Garstecki, P., M.A. Fischbach, and G.M. Whitesides, Design for mixing using bubbles in branched microfluidic channels. Applied Physics Letters, 2005. 86(24): p. 244108. DOI: 10.1063/1.1946902

[4] Weibel, D.B., et al., Torque-actuated valves for microfluidics. Analytical chemistry, 2005. 77(15): p. 4726-4733. DOI: 10.1021/ac048303p

[5] Zeng, S., et al., Microvalve-actuated precise control of individual droplets in microfluidic devices. Lab on a Chip, 2009. 9(10): p. 1340-1343. DOI: 10.1039/B821803J

[6] Kistrup, K., et al., Fabrication and modelling of injection moulded all-polymer capillary microvalves for passive microfluidic control. Journal of Micromechanics and Microengineering, 2014. 24(12): p. 125007. DOI: 10.1088/0960-1317/24/12/125007

[7] Hilber, W., Stimulus-active polymer actuators for next-generation microfluidic devices. Applied Physics A, 2016. 122(8): p. 751. DOI: 10.1007/s00339-016-0258-6

[8] Lee, T.Y., et al., An integrated microfluidic chip for one-step isolation of circulating tumor cells. Sensors and Actuators B: Chemical, 2017. 238: p. 1144-1150. DOI: 10.1016/j.snb.2016.05.163

[9] Zhang, H., et al., Advances in precision micro/nano-electroforming: a state-of-the-art review. Journal of Micromechanics and Microengineering, 2020. 30(10): p. 103002. DOI: 10.1088/1361-6439/aba017

[10] Hou, J., et al., Characterization of manufacturability of microstructures for micro-injection moulding of micro devices using star patterns. Journal of Micromechanics and Microengineering, 2019. 30(2): p. 025001. DOI: 0000-0001-7849-3974

[11] Zhang, H., N. Zhang, and F. Fang, Fabrication of high-performance nickel/graphene oxide composite coatings using ultrasonic-assisted electrodeposition. Ultrasonics Sonochemistry, 2020. 62: p. 104858. DOI: 10.1016/j.ultsonch.2019.104858

[12] Zhang, H., N. Zhang, and F. Fang, Electrodeposition of Nickel/Graphene Oxide Particle Composite Coatings: Effect of Surfactants on Graphene Oxide Dispersion and Coating Performance. Journal of the Electrochemical Society, 2020. 167(16): p. 162501. DOI: 0000-0001-7849-3974

[13] Pina-Estany, J., A. García-Granada, and E. Corull-Massana, Injection moulding of plastic parts with laser textured surfaces with optical applications. Optical Materials, 2018. 79: p. 372-380. DOI: 10.1016/j.optmat.2018.03.049

[14] Zhang, H., et al., Precision replication of microlens arrays using variotherm-assisted microinjection moulding. Precision Engineering, 2021. 67: p. 248-261. DOI: 10.1016/j.precisioneng.2020.09.026

[15] Kim, M., B.-U. Moon, and C.H. Hidrovo, Enhancement of the thermo-mechanical properties of PDMS molds for the hot embossing of PMMA microfluidic devices. Journal of Micromechanics and Microengineering,

2013. 23(9): p. 095024. DOI: 10.1088/0960-1317/23/9/095024

[16] Yu, D., et al., Roll-to-roll manufacturing of micropatterned adhesives by template compression. Materials, 2019. 12(1): p. 97. DOI: 10.3390/ma12010097

[17] Zhang, F. and H.Y. Low, Anisotropic wettability on imprinted hierarchical structures. Langmuir, 2007. 23(14): p. 7793-7798. DOI: 10.1021/la700293y

[18] Waheed, S., et al., 3D printed microfluidic devices: enablers and barriers. Lab on a Chip, 2016. 16(11): p. 1993-2013. DOI: 10.1039/C6LC00284F

[19] Tsao, C.-W. and D.L. DeVoe, Bonding of thermoplastic polymer microfluidics. Microfluidics and nanofluidics, 2009. 6(1): p. 1-16. DOI: 10.1007/s10404-008-0361-x

[20] Mukhopadhyay, S., J. Banerjee, and S. Roy, Effects of liquid viscosity, surface wettability and channel geometry on capillary flow in SU8 based microfluidic devices. International Journal of Adhesion and Adhesives, 2013. 42: p. 30-35. DOI: 10.1016/j.ijadhadh.2012.12.001

[21] Mathur, A., et al., Characterisation of PMMA microfluidic channels and devices fabricated by hot embossing and sealed by direct bonding. Current Applied Physics, 2009. 9(6): p. 1199-1202. DOI: 10.1016/j.cap.2009.01.007

[22] Stroink, T., et al., On-line multidimensional liquid chromatography and capillary electrophoresis systems for peptides and proteins. Journal of Chromatography B, 2005. 817(1): p. 49-66. DOI: 10.1016/j.jchromb.2004.11.057

[23] Guttenberg, Z., et al., Planar chip device for PCR and hybridization with surface acoustic wave pump. Lab on a Chip, 2005. 5(3): p. 308-317. DOI: 10.1039/B412712A

[24] Lu, X., et al., Recent developments in single-cell analysis. Analytica Chimica Acta, 2004. 510(2): p. 127-138. DOI: 10.1016/j.aca.2004.01.014

[25] Watts, P. and S.J. Haswell, The application of micro reactors for organic synthesis. Chemical Society Reviews, 2005. 34(3): p. 235-246. DOI: 10.1039/B313866F

[26] Duffy, D.C., et al., Rapid prototyping of microfluidic systems in poly (dimethylsiloxane). Analytical chemistry, 1998. 70(23): p. 4974-4984. DOI: 10.1021/ac980656z

[27] Unger, M.A., et al., Monolithic microfabricated valves and pumps by multilayer soft lithography. Science, 2000. 288(5463): p. 113-116. DOI: 10.1126/science.288.5463.113

[28] Tsao, C.-W., Polymer microfluidics: Simple, low-cost fabrication process bridging academic lab research to commercialized production. Micromachines, 2016. 7(12): p. 225. DOI: 10.3390/mi7120225

[29] Tan, S.H., et al., Oxygen plasma treatment for reducing hydrophobicity of a sealed polydimethylsiloxane microchannel. Biomicrofluidics, 2010. 4(3): p. 032204. DOI: 10.1063/1.3466882

[30] Shiroma, L.S., et al., Self-regenerating and hybrid irreversible/reversible PDMS microfluidic devices. Scientific reports, 2016. 6: p. 26032. DOI: 10.1038/srep26032

[31] Wlodarczyk, K.L., D.P. Hand, and M.M. Maroto-Valer, Maskless, rapid manufacturing of glass microfluidic devices using a picosecond pulsed laser. Scientific reports, 2019. 9(1): p. 1-13. DOI: 10.1038/srep26032

[32] Knowlton, S., et al., 3D-printed microfluidic chips with patterned, cell-laden hydrogel constructs. Biofabrication, 2016. 8(2): p. 025019. DOI: 10.1088/1758-5090/8/2/025019

[33] Zhang, N., et al., Manufacturing microstructured tool inserts for the production of polymeric microfluidic devices. Journal of Micromechanics and Microengineering, 2015. 25(9): p. 095005. DOI: 10.1088/0960-1317/25/9/095005

[34] Madou, M., Fundamentals of microfabrication: the science of miniaturization. 2nd ed. Boca Raton: CRC; 2002. p. 752. DOI: 10.1201/9781482274004

[35] Schuster, R., et al., Electrochemical micromachining. Science, 2000. 289(5476): p. 98-101. DOI: 10.1126/science.289.5476.98

[36] Wimberger-Friedl, R., Injection molding of sub-(mu) m grating optical elements. Journal of injection molding technology, 2000. 4(2): p. 78. DOI: 217290800

[37] Mekaru, H., et al., Microfabrication by hot embossing and injection molding at LASTI. Microsystem technologies, 2004. 10(10): p. 682-688. DOI: 10.1007/s00542-004-0401-8

[38] Yu, L., et al., Experimental investigation and numerical simulation of injection molding with micro-features. Polymer Engineering & Science, 2002. 42(5): p. 871-888. DOI: 10.1002/pen.10998

[39] Tseng, S.-C., et al., A study of integration of LIGA and M-EDM technology on the microinjection molding of ink-jet printers' nozzle plates. Microsystem technologies, 2005. 12(1-2): p. 116-119. DOI: 10.1007/s00542-005-0014-x

[40] Dunkel, K., et al., Injection-moulded fibre ribbon connectors for parallel optical links fabricated by the LIGA technique. Journal of Micromechanics and Microengineering, 1998. 8(4): p. 301. DOI: 10.1088/0960-1317/8/4/007

[41] Zhang, H., N. Zhang, and F. Fang, Synergistic effect of surfactant and saccharin on dispersion and crystal refinement for electrodeposition of nanocrystalline nickel/graphene oxide composite. Surface and Coatings Technology, 2020. 402: p. 126292. DOI: 10.1016/j.surfcoat.2020.126292

[42] Dai, W., et al., Experiment design and UV-LIGA microfabrication technology to study the fracture toughness of Ni microstructures. Microsystem technologies, 2006. 12(4): p. 306-314. DOI: 10.1007/s00542-005-0056-0

[43] Malek, C.K. and V. Saile, Applications of LIGA technology to precision manufacturing of high-aspect-ratio micro-components and-systems: a review. Microelectronics journal, 2004. 35(2): p. 131-143. DOI: 10.1016/j.mejo.2003.10.003

[44] Munnik, F., et al., High aspect ratio, 3D structuring of photoresist materials by ion beam LIGA. Microelectronic Engineering, 2003. 67: p. 96-103. DOI: 10.1016/S0167-9317(03)00064-9

[45] Worgull, M., M. Heckele, and W. Schomburg, Large-scale hot embossing. Microsystem technologies, 2005. 12(1-2): p. 110-115. DOI: 10.1007/s00542-005-0012-z

[46] Zhao, J., et al., Effects of process parameters on the micro molding process. Polymer Engineering & Science, 2003. 43(9): p. 1542-1554. DOI: doi.org/10.1002/pen.10130

[47] Rötting, O., et al., Polymer microfabrication technologies. Microsystem Technologies, 2002. 8(1): p. 32-36. DOI: 10.1007/s00542-002-0106-9

[48] Ahmmed, K., C. Grambow, and A.-M. Kietzig, Fabrication of micro/nano structures on metals by femtosecond laser micromachining.

Micromachines, 2014. 5(4): p. 1219-1253. DOI: 10.3390/mi5041219

[49] Jensen, M.F., et al., Rapid prototyping of polymer microsystems via excimer laser ablation of polymeric moulds. Lab on a Chip, 2004. 4(4): p. 391-395. DOI: 10.1039/B403037K

[50] Zhang, H., et al., Filling of high aspect ratio micro features of a microfluidic flow cytometer chip using micro injection moulding. Journal of Micromechanics and Microengineering, 2018. 28(7): p. 075005. DOI: 10.1088/1361-6439/aab7bf

[51] Pirskanen, J., et al., Replication of sub-micrometre features using microsystems technology. Plastics, Rubber and Composites, 2013. 34(5-6): p. 222-226. DOI: 10.1179/174328905X64722

[52] Kim, S.-W. and L.-S. Turng, Three-dimensional numerical simulation of injection molding filling of optical lens and multiscale geometry using finite element method. Polymer Engineering & Science, 2006. 46(9): p. 1263-1274. DOI: 10.1002/pen.20585

[53] Kamal, M.R., A.I. Isayev, and S.-J. Liu, Injection molding: Technology and fundamentals. 2009: Hanser Gardner Publications. DOI: 10.3139/9783446433731

[54] Tosello, G., Micro injection molding. 2018: Carl Hanser Verlag GmbH Co KG. DOI: 10.3139/9781569906545

[55] Zhang, H., et al., Precision replication of micro features using micro injection moulding: Process simulation and validation. Materials & Design, 2019. 177: p. 107829. DOI: 10.1016/j.matdes.2019.107829

[56] You, J.B., et al., A doubly cross-linked nano-adhesive for the reliable sealing of flexible microfluidic devices. Lab on a Chip, 2013. 13(7): p. 1266-1272. DOI: 10.1039/C2LC41266G

[57] Zhang, Y., K. Gao, and Y. Fan, Application of a new UV curable adhesive for rapid bonding in thermoplastic-based microfluidics. Micro & Nano Letters, 2019. 14(2): p. 211-214. DOI: 10.1049/mnl.2018.5479

[58] Tsao, C.-W. and W.-C. Syu, Bonding of thermoplastic microfluidics by using dry adhesive tape. RSC Advances, 2020. 10(51): p. 30289-30296. DOI: 10.1039/D0RA05876A

[59] Uba, F.I., et al., High process yield rates of thermoplastic nanofluidic devices using a hybrid thermal assembly technique. Lab on a Chip, 2015. 15(4): p. 1038-1049. DOI: 10.1039/C4LC01254B

[60] Tsao, C., et al., Low temperature bonding of PMMA and COC microfluidic substrates using UV/ozone surface treatment. Lab on a Chip, 2007. 7(4): p. 499-505. DOI: 10.1039/B618901F

[61] Ogilvie, I., et al. Solvent processing of PMMA and COC chips for bonding devices with optical quality surfaces. in 14th international conference on miniaturized systems for chemistry and life sciences. 2010. DOI: 9780979806438

Chapter 4

Microfluidic Flow Sensing Approaches

Liji Huang

Abstract

Precise flow metrology has an increasing demand in many microfluidic related applications. At the scale and scope of interests, Capillary number instead of Reynold number defines the flow characteristics. The interactions between fluid medium and flow channel surface or the surface tension, cavitation, dissolution, and others play critical roles in microfluidic flow metrology. Conventional flow measurement approaches are not sufficient for solving these issues. This chapter will review the currently available products on the market, their microfluidic flow sensing technologies, the technologies with research and development, the major factors impacting flow metrology, and the prospective sensing approaches for future microfluidic flow sensing.

Keywords: Microfluidics, Flow sensing, Thermal time-of-flight sensing, MEMS sensing, Flow metrology, Cavitation, Dissolution, Interface

1. Introduction

Microfluidics is a broad terminology covering various disciplines and scopes while focusing on life science, biochemical and chemical applications. It applies to the devices that process fluids at a dimension below the millimeter scale, and the maximum fluidic volume is within milliliters. Current applications are more related to fluids from microliters to nanoliters in volumes [1, 2]. Two devices in the late 1970s marked the birth of microfluidics. IBM first reported the inkjet printer heads [3] in 1977, and now millions of such units have been shipped worldwide, enabling color printing into every corner of human life. Another device is the micro gas chromatography made on a 5 cm silicon wafer by Stanford University in 1979 [4]. This *lab-on-a-chip* concept has spiked a great enthusiasm and many research activities to miniaturize the analytical instrumentation for wider usage. However, due to its system issues, its progress is less pronounced. Still, progress made on the key components allows its commercialization success with the products now offered by a few companies such as Agilent and Thermo Fisher. The *Organs-on-chips* [5] approaches utilize microfluidic devices to culture living cells for modeling physiological functions of tissues and organs, making microfluidics a unique tool to enrich our understanding of life sciences and to assist the research and assembly of new drugs. Nevertheless, despite tremendous activities in the past 40 years and many applications proven to be feasible, commercialization is limited. Today's microfluidics is yet the well-established one for implementation but excellent academic approaches and science and technology tools [6–11].

The majority of applications of the current microfluidics are for liquid handling other than the gaseous materials. DNA/Gene analysis and point of care disease diagnosis are extensively studied with microfluidic devices [12–15]. The microfluidic devices can integrate both active and passive elements inside the fluidic channels, enabling the polymerase chain reaction devices that help the DNA amplification. The detailed analyses of the DNA samples become possible. Such information is critical for identifying diseases and understanding the origins of the abnormality to the search for possible recovery routes. The other important and advantageous benefits of these microfluidic-based diagnostic devices are fast processing time with small sample volume. These features are combined with today's communication infrastructure, making the remote diagnosis a fascinating scenario. However, current devices are still less sophisticated to acquire the necessary data for the desired tasks. Most of the devices available are based on colorimetric or optical images or limited electrical signals. These data are similar to the analog ones in the electronic age. Additional *digital* sensors will have to be integrated into the microfluidic device for diagnostic quality data acquisition. If the massive deployment of such devices were realized, the foreseeable heavy burdens of the medical systems for the aged societies worldwide would be significantly alleviated, and many lives of human beings could be saved. Drug delivery is another major application for microfluidics [16–18]. In addition to the fluid handling channels and mixers, drug delivery will require a more complicated system that would involve precision metrology, biocompatible carriers, actuation, execution, and feedback. Reliable, low-cost, and commercially available devices will be the keys for future precision drug delivery implementations. The processing of the fluids at a small scale also provides fundamental new tools. It makes it possible to design and fabricate new materials with special functions that the conventional approaches could never succeed. These new materials, including organics, polymeric microparticles, nanostructured materials, and composites, are also the focuses of current microfluidic applications [19].

Microfluidics advancement, on the other hand, greatly relies on the device fabrication technologies of micromachining. The earlier simple passive microfluidic chips having the only microchannels are no longer the mainstream but components of the current devices. A sophisticated microfluidic chip would have both passive structures and active components, which is a challenge for the micromachining process technologies that do not have standard protocols. The multi-discipline features further complicate the availability of process tooling. Fortunately, microfluidics' growth is parallel with the significant advancement in the MEMS and LSI/VLSI IC industry. With ever-improving micromachining device fabrication technologies, the microfluidics once was only viable on a 2″ wafer, and now 8″ and even 12″ wafers are being routinely produced. Many more foundries are available with specialized alternative substrates of glasses, plastics, polymers, and even papers. In recent years, 3D printing, precision micro-injection, laser processing, hot embossing [20–23], and other alternative tools also greatly enriched the variety of microfluidic devices. The progress significantly solves the issues for chemical and bio-compatibility and, in some cases, for commercialization, but the cost to fabricate a desired microfluidic chip is still far from satisfactory. Moreover, the interactions among the device components and the fluids are also likely and sometimes are mandatory, adding additional requirements for better materials and fabrication technologies. Several key components, including microfluidic channels, microvalves, micropumps, needles, mixers, and sensors, are considered the necessary ones for the desired microfluidic chip or system. These relatively complicated components and the substrate make the process compatibility with the electronics a dilemma. Therefore, package, interface, and system design will become critical for the device's final footprint, manufacturability, and successful deployment.

The inkjet printer head that handles the ink droplets remains an outstanding example of a successful microfluidic application. The envisioned microfluidic future in life science and others are still missing a bridge, mostly from the ease of reach and cost-effectiveness [9, 24–26]. The research data for the current microfluidic market have excluded the inkjet applications, addressing only the diagnostic devices and pharmaceutical and life science tools [27]. Nevertheless, by comparing the market reports from the same market research firm issued a decade ago to the current data, one could find that even the most optimal old forecast has nothing to beat the real growth. On the other hand, today's multi-billion dollar market and the double digital growth predicted by various market research firms are more from the companies making the system level products but not the direct values of the key components. These data for the value-added systems are, in a sense, could deceit the current research focus on components. While the system level products enable various applications, the lack of a miniaturized, standalone, performance dramatizing, and cost-effective device would not maintain the expected or envisioned phenomenal growth. In this chapter, standalone flow sensor products for microfluidics will be discussed, including the technologies, standards, factors that will impact the performance, integrations, and manufacturability or scalability.

Microfluidic studies have covered a huge spectrum of processes. For this chapter's limited space, only continuous flow sensing technologies are discussed with applicable pulsed flow features. Droplet flow, nanofluidic flow, microfluidic manipulation or handling, and biological and chemical-related flow phenomena will not be addressed.

2. Microfluidic flow sensing technologies

Microfluidic sensors are critical components for a complete system. Many research works on sensors have been dedicated to the biomedical and chemical sensing development based on electrochemical, optical, mass, or magnetic sensing principles. Electrochemical sensors are mostly studied and often composed of several electrodes that are easy to fabricate together with the microchannels. This limited integration with a simple configuration allows a fast response with reasonably good sensitivity and enables multiple reagents on a single microfluidic chip. The electrode embedded inside a microfluidic channel can also be used for cell counting and estimate the flow volume. With the assistance of a miniaturized LED, pH measurement could be achieved as well [28–31]. However, many of the proposed biosensors or chemical sensors are very specific, and most are research-orientated, as being determined by the catalytic or affinity properties of the biological recognition agent in a particular study and the sensor itself requires a sophisticated electronic system for readout or analysis. The standalone or large scale commercial applications are yet to emerge.

Flow meters using traditional thermal capillary and Coriolis measurement are commercially available for micro flow measurement before the microfluidics being intensively studied. Researches on microfluidic flow sensing approaches are for miniaturized, cost-effective, and integrable products. In this scope, both flow and pressure sensors have been extensively studied [28]. In some cases, the differential pressure sensor can be used for flow measurement. Flow measurement is one of the most important factors in microfluidic handling for data analysis and precise system control. Without the knowledge of the fluid quantity in the process, analytical results would not be easy to establish the needed and convincing statistics. The conventional flow sensors might be the first commercially available standalone sensing products for microfluidics. The technologies are still limited, and their package formality is bulky and far off the cost target for the desired microfluidic system. Many studies proposed integrating flow sensors into the microfluidic system. However, there are still many factors that impact data acquisition. The existing sensor products on the market also have some unsolved reliability issues in applications. The commercialization route to a well-performed and cost-effective sensor is yet to be demonstrated.

The available flow sensors applied to microfluidics are classified as thermal and non-thermal sensors [1]. Thermal flow sensors have been applied to small flow measurement for both gas and liquid before the microfluidic concept emerged. Therefore thermal flow sensors are mostly studied and applied in microfluidic applications, and products with various thermal sensing principles are commercially available. Coriolis microfluidic sensor is a non-thermal sensor, and it has an even higher cost. Other "non-thermal" flow sensors are mostly at the research stages. Before the form factor, cost, and reliability issues can be solved, large scale applications are still not possible.

2.1 References and standards

For the traditional flow sensors, the metrology characteristics will hardly enable a self-calibration. Therefore, a primary standard or a reference defined by an international norm governs the manufacture of a flow sensing product with specific sensing technology. The same should then apply to microfluidics. Demanding to establish an international standard for microfluidics has long been proposed [32, 33]. Still, only in recent years, an international microfluidic association has been established, and an international standard (ISO) working committee has been organized with a serial of workshops [34]. It has been proposed that the new ISO standard for the microfluidic shall be having four sub-standards, including *flow control* that addresses the key components of valves, pumps, and sensors for the system; *Interfacing* that is to standardize the connectors and other interfaces; *modularity* that will regulate the integration and *testing methods* that will define the methodology of the metrology and other related testing issues. Such a task is still at an earlier stage, and additional time will be need before the standards become available.

Several efforts to establish a primary standard or a traceable reference system for flow metrology in microfluidics applications have been made in the past years [35–37]. The widely adapted primary standards are the gravimetric and volumetric principle. The comparison of such standards among different European national metrology institutes indicated an uncertainty (k = 2) ranging from 0.05 to 6% for the flow rate ranges of 17 nl/min to 167 ml/min. Still, most of them can have uncertainties within 0.1% [35]. In the reported reference system, the flow generation is critical and requires a stable flow system in the sub microliter per minute flow. Other special effects such as evaporation must be considered, especially when approaching nanoliter per minute flowrates. Degassing through the system and preventing external vibration, and testing environmental control are also critical to ensure the measurement is repeatable and accurate. The flow generators used in these institutes include metallic bellow, precision syringe pump, and gear pump. Because the gravimetric measurement is achieved with high precision balances, the system is a uni-directional open loop. For sub microliter per minute flowrates, laser interferometry has been used as an alternative precise reference for the desired accuracy [38]. However, for high volume applications, a faster closed-loop calibration would be preferred. It could also result in good accuracy using a gear pump and high precision Coriolis meter with an accuracy of ±0.2% as the reference standard [39].

For almost all flowrate ranges in microfluidics, the Reynold numbers are within 1000, indicating that the flow of interests is within the laminar flow regime. Therefore in a desired large dynamic range, the flow profile would not be the same at the different flowrates, which adds complexity to maintain the measurement accuracy. Meanwhile, the flow channels are small in micrometer dimensions. The interfaces between fluid and channel wall become pronounced, which differ from those described for laminar flow by Moody Diagram in the classic fluid dynamics. Besides, cavitation would play a critical role, and dissolution will also contribute to metrology. These are among the new challenges for the on-going metrology standards for microfluidics.

2.2 Thermal flow sensing

The thermal mass flow measurement using calorimetric capillary sensors has been used to measure a very low flow to nanoliter per minute for quite a long time [40]. The sensors are composed of thin metal wires winded outside the wall of a tiny tube of a micrometer in diameter. The tube is usually made of thermally conductive materials such as stainless steel or fused silica. These sensors normally require a higher power to ensure the heat transfer resulting in a small dynamic measurement range and a low accuracy towards the low measurement end. The required manufacture process makes these sensors very costly without being able to be volume produced. Integration of such a sensor into a microfluidic system would be unlikely. In the following discussions, only micromachined sensors will be addressed. The micromachined sensors are mostly made on silicon or glass substrate. A microheater and plural numbers of sensing elements are deposited on a membrane structure, and the air or gas-filled cavity below the membrane provides the desired thermal isolation. The tiny sensing elements enable a fast response time. The membrane is frequently made with silicon nitride or silicon nitride and oxide combination. The sensing elements can be metals with a large temperature coefficient such as platinum, nickel, tungsten, or in the case for the process compatibility, doped polycrystalline silicon is used instead. The micromachined thermal flow sensors' structure has no moving parts, and the surface can be treated with various passivation and post-process coating for better reliability. The micromachining process for the flow sensors is well established today. Most MEMS foundries have the necessary equipment for manufacturing such sensors, which allows a very favorable cost and makes it possible for high volume applications. The first micromachined thermal flow sensor made for microfluid is used in micro gas chromatography [4]. It is for gaseous flow and not a standalone product and only manufactured in a minimal quantity as the OEM product. The commercially available micromachined thermal microfluidic flow sensors for liquid were incepted in the last decade. These commercial products utilize different thermal sensing principles [41–43] that cover the three major technologies with thermal calorimetry, anemometry, and thermal time-of-flight approaches. There are some research activities on other thermal flow sensing designs, such as thermal capacitive utilizing the temperature dependence of dielectric constants, [44] and temperature dependence of the PN-junction in a diode [45]. The measurement scheme of flowrate with these alternative thermal sensing designs could also be classified into the above three thermal sensing principles. **Figure 1** is the graphic illustration of these three measurement principles for the typical micromachined thermal flow sensors on a silicon substrate.

2.2.1 Calorimetric flow sensing

The majority of the current micromachined commercial thermal flow sensors are utilizing the calorimetric principle. Most successful applications are for gaseous

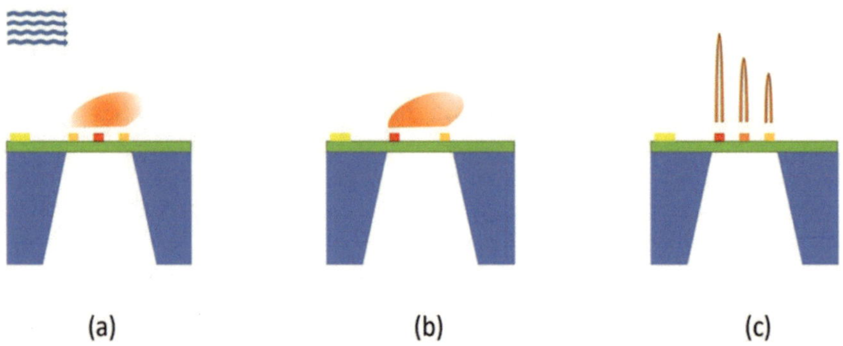

Figure 1.
Graphic illustration of the micromachined thermal flow sensors (on silicon) with the flow sensing principles: (a) calorimetry; (b) anemometry and (c) thermal time-of-flight.

fluids, of which the automotive airflow sensors for fuel control are the dominant application. The structure showed in **Figure 1(a)** is a typical one for a micromachined calorimetric mass flow sensor on which a microheater is placed at the center of the membrane. Two temperature sensors are made symmetrically at the up and downstream of the microheater. These two temperature sensors can be simple resistors of identical resistance values or identical thermal-piles. There are a variety of approaches to realize data acquisition. The commonly used ones are either to keep the microheater at a constant heating power or to maintain a constant temperature from the up and downstream sensor and then measure the heat transfer or temperature differences between the measurements of the up and downstream sensors as the flowing fluid will take away the heat from the microheater resulting in a heat redistribution. By calibration, such heat transfer can be correlated to the mass flowrate of the fluids. In this approach, the measurement is susceptible to low flowrate. As its nomenclature indicates, the measurement is dependent on the fluidic thermal properties of thermal capacitance and thermal conductivity. The thermal sensing using the resistor-based microheater and resistor sensing has the intrinsic temperature effects associated with the environmental conditions, which need to be compensated for better accuracy. For this purpose, another sensor placed on the substrate (the yellow element shown in **Figure 1**) is used to gauge the environmental temperature and correct the resistance value due to environmental temperature variations. The detailed theoretical interpretation and governing physics can be found in the literature as well as the international standard [1, 46].

The major challenge of applying the micromachined thermal sensor to meter microfluidic is the package. For the gaseous sensors, the membrane often has openings that balance the surface's fluidic pressure against the membrane deformation. The change of the membrane position will greatly impact the measurement as the sensor position will be significantly altered with membrane flatness changes. However, for microfluidic measurement, the opening will be detrimental once the liquid-filled up the cavity underneath the membrane. Therefore, the commercially available approach [41] for the package is to have the sensor placed outside the channel with the sensor's surface close to the outer channel wall. Therefore, the channel will need to be thin enough and have good thermal conductivity for heat transfer effectiveness. One of the selections of the channel is a fused silica tube. As the membrane that supports the sensing element is typical with a thickness of 1 micrometer, attach the think tube to the sensor is a very tedious process with a high cost. In addition, compared to the applications for gaseous fluids, the thermal wall of the fused silica also reduces the sensitivity of the sensor, leading to a significantly smaller measurement dynamic range (<50: 1), which is certainly not desired for microfluidic applications. The commercially available calorimetric microfluidic sensors offer a typic <40:1 dynamic range with the lowest detection flowrate of 7.5 nL/min and the best accuracy of ±5% of reading at the full scale. There are also concerns about the constant thermal power at the channel's specific area during the measurement in practical applications. This will be discussed later in detail.

2.2.2 Anemometric flow sensing

The first micromachined thermal flow sensor on silicon is made with the anemometric flow sensing principle [47]. Thermal anemometry is also known as energy dissipative sensing, and its measurement scheme is relatively simple, as shown in **Figure 1(b)**. Only one sensing element is placed downstream. Alternatively, the sensing element can also be placed upstream, as the measurement of the fluidic flowrate is only from the microheater (a sensing element). The temperature sensor is used as a fluid temperature reference. Therefore, instead of measuring the fluidic flow-induced changes of the temperature profile at the centralized microheater with calorimetry, the anemometry measures the heat loss due to the forced convection. In this case, with the supporting control circuitry, adjusting the microheater power will allow the measurement to be much easier for higher flowrates. Simultaneously, the sensitivity at low flow will be lower compared to the sensing principle of calorimetry. Another character of the anemometry is that its correlation with the fluidic thermal properties has a larger nonlinear effect resulting in the difficulties to apply a constant fluidic conversion factor for correction of the flowrate data when the measured fluid has different thermal properties from those of the calibration fluid. For the same reason, the temperature compensation scheme for the anemometry is more complicated than that for the calorimetry.

One commercially available anemometric microfluidic flow sensor, per the structure described in the company's webpage, [43] also takes the package approach similar to the earlier mentioned one of the calorimetric microfluidic sensors. The sensor is placed at the outer wall of a thermally conductive fine quartz glass tube by machining the tube surface into a smooth flat. Instead of a single micromachined sensing chip, two chips are used. A special glue was applied to attach the chip to the quartz tube's flat surface, forming a close contact for the required heat transfer. The heater chip and temperature chip are separated at a certain distance forming the configuration of an anemometer. The heat transfer needed for the measurement provided by the sensor is achieved via thermal diffusion. These package approaches are also similar to the traditional capillary thermal mass flow sensors, where the hot wires are winded onto the surface of a special stainless tube. However, the micromachined sensor will have a much lower heating temperature than those by the capillary sensor. Because of the heat diffusion, control the heat for the low flowrate measurement would be very challenging, resulting in a small dynamic range and large measurement errors (full-scale error rate) towards the low detection limit. The current offered anemometric microfluidic flow meter has a guaranteed dynamic range of 50:1 with the lowest detectable flowrate of 100 μL/min and the best accuracy of ±5% of reading.

2.2.3 Thermal time-of-flight flow sensing

Both the calorimetric and anemometric flow sensors require a calibration of the real fluid for the desired precision or metrological accuracy, as the fluidic properties will have a nonlinear response in the full dynamic range. The limited dynamic range and the accuracy would not be desirable for the precision requirements for many microfluidic applications such as drug infusion. Also, these flow sensing products could only provide mass flowrate measurements. The microfluidic applications would appreciate additional fluidic information such as fluidic concentration, physical or even chemical properties of the fluids at the same time. To this end, thermal time-of-flight sensing technology offers much of the competitive advantages. The thermal time-of-flight sensing concept can be traced back to the late 1940s [48] and has been an interest in many subsequent research works [49–51]. The thermal time-of-flight sensing measures the heat transfer transient time as well as the responses at each sensing element. Several sensing elements can be placed downstream of the microheater. Consequently, this approach can measure additional parameters other than the flowrates [52]. The sensor works with a thermal pulse or modulated thermal wave signals. Compared to calorimetry or anemometry, the transient time-domain data are much more immutable to the background interferences. Despite the advantages, a commercially available thermal time-of-flight flow meter is not seen until the past decade [42]. One reason could be that the microheater must possess a mass as small as possible for the needed thermal response to enable the measurement scheme. In the traditional approach, such a tiny wire is extremely vulnerable for reliability in actual applications. On the other hand, pure time-of-flight will only measure the flow velocity. In contrast, the other parameters require advanced and complicated electronics that are not readily accessible until recent years. Nevertheless, the sensor build and package limitation will still lead to a non-pure time-of-flight, and calibration will be required to remove those effects. On the other hand, these effects can also be used to provide additional fluidic information. For the microfluidic applications, the microheater is driven with a modulated microheater, the constant heating spot in the flow channel is therefore eliminated. The sensor outputs flow velocity as well as fluidic mass flowrate and the additional data of the fluidic properties, making the thermal time-of-flight technology an ideal approach for the desired microfluidic flow measurement applications.

Figure 2 shows a typical structure of a micromachined thermal time-of-flight sensor chip [53]. The micromachined process has a wide spectrum of materials selection to allow the sensor with excellent thermal isolation while not sacrificing reliability. This is particularly important for the thermal time-of-flight sensing that requires a super-fast thermal response. The blue materials showed in **Figure 2(b)** can be silicon or glass substrate. The gray colored block will be for thermal isolation. For example, a 10 ~ 15 μm parylene conformal layer will provide the properties of the good material of stiffness and robustness for the application. The green-colored materials need to have good thermal conductivity while excellent surface passivation for reliability. Ideal materials include multi-layered silicon nitride or silicon carbide. Underneath the microheater and sensing elements, a cavity will enhance the thermal performance of the sensor chip. The brown-colored elements are for microheater and sensing elements. One sensing element is placed directly on the substrate to measure the environmental temperature that provides the compensation of the microheater's temperature performance and control. In the photo shown in **Figure 2(a)**, the central element has another sensing element at the proximity of the microheater, which is used to fine-tune the microheater temperature or power with those other physical properties such as thermal conductivity can be precisely acquired.

The heat transfer in the thermal time-of-flight configuration is measured by the temperature T with time t in the flow direction x of non-uniform temperature distribution, determined by the flow medium's conductivity and diffusivity. The working principle can be expressed as below, [52].

$$\frac{\partial T}{\partial t} = \frac{k}{\rho c}\left(\frac{\partial^2 T}{\partial x^2}\right) - V_x\left(\frac{\partial T}{\partial x}\right) + \frac{Q}{\rho c} \tag{1}$$

Figure 2.
Example of a micromachined thermal time-of-flight sensing chip: (a) optical photo of the chip, top view; (b) cross-section schematic.

Where Q is the heat generated by the microheater that is normally modulated with a square or sine wave, k is the thermal conductivity, ρ is the density, c is the heat capacity of the flow medium, and V_x is the fluid velocity. With one dimensional approximate, the fluid velocity, V_x, between the microheater and the sensing element would be determined by fluid thermal conductivity k, thermal diffusivity α, and the modulated heat Q, [54].

$$T = \frac{Q}{4\pi kt}\exp\left(-\frac{(x - V_x t)^2}{4\alpha t}\right) \tag{2}$$

Therefore, if the sensor only has a microheater and a sensing element pair, the measurement will still be dependent on the flow medium properties. The microheater and the sensing elements all have the fluidic dependent response that needs to be removed for the complicated fluids. Simple calibration with the conventional fluid can be applied for the fluid measurement without losing the metrology accuracy. A micro-machining process' advantages allow placing multiple sensing elements on the same chip without adding any cost that makes it possible to have the measurement independent of the fluidic properties. The thermal time-of-flight will not be a simple flow velocity measurement. The measured changes in the amplitude are directly proportional to the heat transfer between the microheater and the sensing elements that will provide the mass flowrate similar to the calorimetric or anemometric approach per the data acquisition process. The time-domain data yield additional information, which allows the acquisition of additional fluidic thermal dynamic properties such as thermal conductivity and specific heat. In the microfluidic flow measurement, the liquid is generally non-compressible. The pressure effects of compressibility can be considered secondary. Compared to the gaseous fluids, liquid has a much large heat capacitance making the sensing element resistance-related temperature effects less pronounced. And most importantly, with the multiple sensing elements on a single chip, the measurement dynamic range can be substantially extended. A practical 7500:1 dynamic range can be achieved with two or three pairs of sensing elements.

2.3 Coriolis flow sensing

Coriolis mass flow sensing principle has been well documented, and the first commercial product was introduced to the market in 1977 by Micro Motion. It is a true mass flow sensing technology with very high precision by utilizing an exciting tube which fluid is flowing through, and the tube oscillates artificially. The changes of the tube oscillation in time and space are a direct measure of the mass flow. One advantage of the Coriolis sensing approach is that the fluid density can be simultaneously determined from the oscillation frequency of the measuring tube. But it also requires a minimum density of fluid such that the resolution of the oscillation can be registered. It also suffers a high-pressure loss and a smaller dynamic range. The high cost of the measuring tube manufacture sparked the attempt with micromachining, and the first research paper was published in 1997 [55]. With the micromachining process, the MEMS Coriolis mass flow sensor can be well applied for microfluidic flow measurements. The commercially available Coriolis meters sensors via micromachining either consists of a silicon microtube via silicon wafer fuse bonding and an integrated temperature sensor [56] or a silicon-rich silicon nitride tube coupled with a strain gauge readout [57, 58]. The micromachined Coriolis sensor using silicon nitride tube has a thin tube wall of about 1.2 μm and is much lighter than the silicon tube. Hence, the light-weighted tube would have a smaller mass than the fluid it measures that simplify the package, and leads to the possibility to measure the fluids at ambient pressure. The demonstrated Coriolis sensor could measure liquid mass flow, density, and temperature (if a temperature sensor is integrated) simultaneously. Another advantage for the MEMS Coriolis mass flow sensor is that it usually operates at a much higher resonant frequency with substantially less vibratory influences from the environments than those for the traditional Coriolis mass flow technology.

Like the MEMS thermal mass flow sensors, the micro Coriolis mass flow sensor also requires clean fluid. Even fine particles can damage or clog the sensor, considering the measuring tube's tiny channel. Besides, the sensor will not function well in liquid with high viscosities and liquid with chemical reactions. The high-speed liquid flow may also alter the performance of the sensor unless bypass configuration is applied. The superior true mass flow accuracy of a Coriolis sensor is overshadowed by its footprint, complication in the package, and cost in the manufacturing process that diminishes high volume and/or disposable applications, which would be a necessity in some microfluidic applications for cross-contamination prevention. The fluidic property independence measurement characteristics also limit its measurements only for flowrate and density. Other fluidic property measurements will require integrating additional sensing elements, further enlarging the sensor footprint, indicating an even higher cost for the final product manufacture. Therefore, although the micromachined Coriolis sensor's demonstration has been over two decades, the applications are still very limited.

2.4 Other microfluidic flow sensing technologies

Flow sensors are likely the ones that can be made with the most versatile technologies and are vastly selectable to the applications. More than twenty different physical measurement principles are commercially available on the market for flow metrology. However, for microfluidics, the options are limited. Other than the commercially available thermal mass flow sensors and Coriolis flow sensors, micromachining advancement offers opportunities for many studies with a wide spectrum of technologies applied for microfluidic flow sensing. But commercialization of many of those is still in question. Some selected researches micromachined flow sensing technologies are discussed below.

2.4.1 Acoustic microfluidic flow sensing

In addition to micromachined thermal and Coriolis sensors, micromachined ultrasonic sensors are also commercially available. Ultrasonic flow sensing is one of the well-documented technology for flow metrology with high accuracy. By measuring the time differences of the ultrasonic signals propagating in the opposite direction of a fluidic medium, the flow speed can be accurately measured. Therefore, it has the advantage of a pure velocity measurement independent of the fluidic properties. As a sound propagation, it will not require direct contact with the fluids that it measures, or it is non-invasive, which is very attractive for microfluidic related medical applications. However, it will normally require dual transducers placed in opposite directions or at a certain angle with respect to a reflector. This prevents the reduction in footprint and cost. For the microfluidic applications, its signals reduce significantly at the low flow speed, and it is also very sensitive to the fluids where cavitation or dissolution may exist. The current commercially available ultrasonic flow meters for microfluidics have about a 50:1 dynamic range and a detection limit of 300 μL/min [59]. Some research indicated that the ultrasonic transducer could be integrated into the microfluidic channel, but the capability for flow metrology is yet to be demonstrated [60].

Acoustic device applications in microfluidics are mostly for fluid handling, and surface acoustic wave (SAW) sensing and actuation is another approach that can be integrated into the microfluidic channels [61]. Some efforts were also made to measure the flowrate with the SAW devices. It has been reported that a micromachined interdigital transducer (IDT) could direct the fluidic droplets via the excited acoustic streaming that is fast and material independent [62]. In another study, [63] a SAW sensor with dual symmetrical IDTs made on a 30 mm by 30 mm square quartz crystal substrate was used to measure the flowrate in a designed channel by recording the delay times and the corresponding frequencies. A close to a linear correlation between the phase shift from the delay time and flowrate was established. The SAW sensors can be independent of the fluidic properties; however, they require a much larger footprint, and temperature compensation is also complicated compared to thermal sensing approaches.

2.4.2 Differential pressure microfluidic flow sensing

Measurement of flowrate with differential pressure is one of the oldest flow sensing technologies. Micromachined differential pressure sensors have been well established and are widely available on the market at a very low cost. Most sensors are made on a silicon nitride membrane or diaphragm with the piezoresistive sensing elements at the edges of the membrane or with a capacitance measuring principle for the low differential pressures [64, 65]. The advantages of a differential pressure sensor for flow measurement are lower power consumption and relatively easy installation with fewer effects on the flow conditioning. They are also independent of the fluidic properties. The microfluidic flow regime is purely laminar, and the pressure loss is linear with the flow velocity. However, limited by its sensitivity, the measurement dynamic range of a differential pressure sensor is normally small. In particular, for microfluidic applications, the pressure drop at a tiny distance may not even generate enough sensitivity for the measurement. The dependence of the microfluid's pressure loss on the dynamic viscosity also requires a temperature sensor at the proximity for the needed compensation. Other phenomena such as cavitation or multi-phase flow will have a big impact on the measurement of the pressure and hence the accuracy of the deduced flowrate.

Flow measurement with drag force is an alternative pressure-related flow sensing approach. Due to the size restriction, such a sensor does not favor being placed inside the microfluidic channel. However, in an ideally integrated microfluidic system, there will be valves and other actuators. The drag force-sensing approach could be combined with the actuation parts in the system. A typical drag force sensor is to utilize a cantilever or a diaphragm [66]. The mechanical deflection can be read out with an optical microscope or photodiode. Another approach to measuring the deflection is to utilize the piezoresistive or piezoelectric elements embedded at the positions where maximal deformation could occur at the designed cantilever or diaphragm. To increase the measurement sensitivity, the Fabry Perot spectrum's fringe shift was used to measure the cantilever movement correlated flowrate, which, however, complicated the data acquisition and limited the package options [67]. The materials used to make the micro-cantilevers are silicon nitride, SU8, and polydimethylsiloxane (PDMS). An integrated micro-cantilever inside the microfluidic channel via the microfluidic favorable PDMS process achieved a capability of detecting 200 μL/min flowrate but only have a small 5:1 dynamic range [68]. Most of the micro-cantilevers measure microliter per minute flowrate, even though nanoliter per minutes sensitivity was reported, but the required optical readout often makes the fine readings and subsequent digitization a challenge [69]. While piezoresistive or piezoelectric configuration is more preferred as no optical assistance in readout will be needed. On the other hand, as the piezoelectric cannot detect a static flow, piezoresistive is considered a better choice. The cantilever sensors are more sensitive than diaphragm sensors, but there are still concerns for their reliability and repeatability per the moving cantilever. The sensitivity of these sensors also requires meaningful pressure or critical mass to activate the deformation of the cantilever or diaphragm. Such pressure is not necessarily existing in the microfluid subject to measurement.

2.4.3 Microwave microfluidic flow sensing

The non-invasive approach is always preferred in microfluidic applications, for which life science is the major focus. A microwave microfluidic flow sensor is reported [70] to achieve a large dynamic range of 1-300 μL/min with a high resolution of 1 μL/min. The detection of the flowrate with the microwave is via the measurement of a membrane that was a part of a microfluidic channel and on which the fluid is flowing over, causing the deflection of the membrane. Therefore, it could also be a type of differential pressure sensing. The measurement element is the microwave resonator that detects the effective capacitance because of the changes in the deflected thin membrane's effective permittivity due to the channel pressure changes by the flowing fluid at different flowrates. This configuration is much easier to be packaged with the microfluidic channel, and it is a true noncontact detection that can be miniaturized compared to the optical assisted readout. The microwave flow sensor is consisting of two critical components. One is the microfluidic channel with the membrane that was micromachined with PDMS soft lithography and replica molding. PDMS is a preferred material for microfluidics for its compatibility, and more importantly, it is transparent to microwave with a low loss. The membrane is about 1.5 to 3 mm in diameter and 100 μm in thickness, strong enough to hold the fluidic pressure inside the microfluidic channel. Simultaneously, it is thin enough for the sensitivity of the resonator function needed for the measurements. The second component of the flow sensor is the microwave resonator, designed into an open-ended half-wavelength ring resonator with a microstrip structure on a high-performance microwave substrate made of a 35 μm copper layer on top and bottom surfaces. The resonator operated at a 4 GHz resonant frequency. The fabrication is via the cost-effective conventional printed circuitry board processing. However, the integration with the microchannel made strong application dependence and difficulties in controlling the cost. Also, the metrological performance of this sensor was not well documented.

2.4.4 Optical microfluidic flow sensing

Optical flow sensing is attractive to the microfluidic application for its non-invasive and high accuracy features. Laboratory flow measurements such as particle image velocimetry, infrared thermal velocimetry, and laser interferometry are reported for microfluidic metrology studies [71–74]. These optical technologies are all having complicated bulk settings and require the microfluidic channels to be optically transparent. While the miniaturization efforts continue to focus on microfluidics, optofluidics is now a dedicated field for the studies of the combined optics and microfluidics with targeted miniaturized optical integration sensing functions into a single microfluidic chip. In a microscale optical flow sensor report, [75] an optical fiber structure was fabricated in the form of a drag force cantilever to measure the microfluidic flow. A stripped single-mode optical fiber was positioned across a microfluidic channel and aligned with a multi-mode fiber receiver. The microfluidic flow in the perpendicular directions will displace the fiber cantilever tip, causing the light intensity change at the aligned receiver. The reported sensor had achieved a measurement dynamic range over 60:1 and a minimal detection of 7 μL/min. However, the making of the sensor would be quite complicated with the fiber alignment, and direct contact of the flowing fluid with the fiber cantilever is also required. In another report using the optical approach for flow sensing, miniaturized fluorescence sensing is attempted for micro molecular tagging velocimetry in microfluidics [76], but these methods are not cost-effective and yet to reach the small footprint.

In an alternative optical sensing approach, [77] a collimated light beam was employed to excite the surface plasmon resonance at a gold film on top of a polymethyl methacrylate (PMMA) microfluidic channel. The fluidic flow will cause the temperature redistribution inside the microfluidic channel, which alters the refractive index above the metal film. The refractive index is inversely proportional to the temperature. By acquiring and analyzing the image of the excited surface plasmon, the flowrate could be measured. However, since surface plasmon resonance is very sensitive to temperature, and the response is nonlinear, a full functional measurement scheme and affirmation of metrology parameters will need additional efforts.

2.4.5 Impedance based microfluidic flow sensing

The impedance flow sensing principle is also a topic in the studies for microfluidics. The electrical impedance flow measurement has the advantage of simplicity. The configuration has fewer requirements for environmental parameter compensation and can be applied to a wide range of fluids. A cascade finger structure of the electrical impedance sensor could help the measurement accuracy of pulsed flow. However, the electrical impedance measurement is strongly dependent on the fluid properties and is only applicable to conductive fluids. In a report [78] of an electrical impedance microfluidic flow sensor, the simple two surface electrodes are embedded inside a microfluidic channel. An alternating current signal was applied across the microfluidic channel. The fluid is equivalent to a diffuse layer capacitance impedance or the parallel capacitance impedance, and the electrode forms the serial capacitance impedance with the fluid. By optimizing the applied voltage frequency, the measured impedance can be well correlated to the flowrate. The reported data achieved a 50 nL/min detection limit and about 10:1 dynamic range.

In another approach, the measurement of the magnetic impedance of a hair microfluidic flow sensor offers the ultra-low-power option [79]. The sensor was made by depositing a giant magnetoimpedance (GMI) layer on top of a glass substrate. A PMMA master pillar mold was then applied to the pre-formed magnetic nanocomposite of permanent magnetic nanowire and PDMS mixture on the GMI layer. The formed flow sensor was placed inside the microfluidic channel. When the fluid flows through the pillars, the flow will force the pillars to bend, resulting in the change of the magnetic field sensed by the GMI layer and output the signals that can be correlated to the flowrate. The results showed a measurement of the water flow speed up to 7.8 mm/sec and a resolution of 15 μm/sec with a typical power of 31.6 μW. The study also indicated that by optimizing the parameters, the power could be further lowered to about 80 nW.

3. Factors impacting the microfluidic flow sensing

The flow metering at the microfluidic scale is quite different from those in a large pipeline. Many factors that may be trivial in the conventional fluidic dynamics become critical for microfluidic metrology. In this section, some critical factors are discussed.

3.1 Microfluidic channel and fluid interactions

In the classic fluidic dynamics, the Moody chart indicates that at laminar flow, the friction factor is inversely proportional to Reynolds number where only viscosity of the fluid plays the role and diffusion is normally not in consideration. In the dimension of a microfluidic channel, the surface area relative to the volume is dramatically larger than those in a large pipe. For the flow speed of interests, factors such as surface tension and diffusion are all having their critical contributions to the microfluidic flow metrology. The capillary number then would be much more important than the Reynolds number [80]. Besides, the majority of microfluidic processes are water-based. Water has a molecular size of about 0.27 nanometer, and it is dipolar in nature. Water interaction with the solid surface is inevitable, and such interaction will be pronounced as interaction will involve a significant portion of the total volume of the microfluidics. Most of the solid surfaces at the microscale would be imperfections that are full of defects with dimensions larger than the water molecule. Water viscosity is also very sensitive to temperature in the applicable ranges. These effects will be even more pronounced in the biological fluid case where the electrolyte is often present as the chemical state of the surface would be altered, either by ionization of covalently bound surface groups or by ion adsorption [81]. Hence, to ensure the accuracy in the flow measurement for microfluidics, the interactions between fluid and solid channel surface must be considered, especially for the long term repeatability, reproducibility, or reliability.

The detailed studies on the fluidic handling and flowrates impacted by the fluid and microchannel interactions are not well documented. However, in a few reports on the long-term stability of the commercially available calorimetric flow sensors for microfluidics, it was reported that the measurement accuracy tended to have a time-dependence. The long-term *drift* was always towards negative directions with a more pronounced deviation at the full-scale flowrate. For example, one report [82] tested the reproducibility of several commercial calorimetric flow sensors of the identical model for the time dependence in water. It was found that although the sensor to sensor performance was inconsistent, the accuracies of all sensors *drifted* towards negative with time, with −25% deviations at the full-scale flowrate in about 5 months. Although the report did not speculate the reasons for the deviations, this phenomenon could be a direct reflection of the water interactions with the microfluidic channel walls. The sensor chip was fixed to a fine tube with a machined flat surface in the product package. It would difficult to guarantee the consistency of such attachment. The heat transfer was from a microheater with a constant heat diffusion at a fixed glass wall area. This area with a constant heat might promote the interaction between water and any defective sites on the inner channel surface, forming an interface with water-filled pinholes that could percolate laterally, reducing the thermal responses because of the wetted surface condition compared to the dry one at the calibration. Hence the measurement would be gradually moving towards the values with negative deviations.

3.2 Microfluidic cavitation

Cavitation is often known as a detrimental phenomenon in high-speed flows that leads to mechanical damages at the flow path. However, it can also be utilized for industrial processing in classic fluidic dynamics. In microfluidics, cavitation inception is via the diffusion of dissolved gas into the available nuclei. It can occur even at a pure Stokes flow, but the cavitating flow will not normally lead to mechanical defectiveness due to the relatively low energy release, but it can dramatically generate the local flow speed spike. Cavitation has become a growing research topic in microfluidics. It is not only because the cavitation flow is inevitable in many applicable microfluidic flow conditions, but it can also be employed as a tool for microfluidic manipulation such as pumping and mixing via the control of cavitation size alternation. The cavitation can harvest and release energy upon collapse in the microfluidic process. The removal of cavitation can be done with properly designed materials for the microfluidic channels [83–86].

The cavitation presence will greatly impact the measurement reproducibility or accuracy for any flow sensors regardless of the measurement principles. The calibration setup for a microfluidic meter normally requires a degassing device in serial to the calibration line, and degassing is always performed before the start of calibration [39]. The cavitating flow is in fact a two-phase flow. Therefore when a flow sensor calibrated at a cavitation-free condition is applied to measure a cavitating flow, the measurement deviations will be inevitable. The current tools of the cavitation studies are visualization approaches such as colorimetry or via high-speed camera for which a transparent flow channel will be required to collect the data. However, in practical applications, the channels are often opaque. Therefore, it is of interest to have additional measurement approaches that can alert *in situ* when the cavitation is present in microfluidics. In some applications such as drug delivery, the infusion with cavitating fluids into a human body will be very harmful, not just for the uncertainties in totalized delivered drugs. Of the flow sensing technologies discussed above, the thermal sensing approach could be a viable tool to detect cavitation while correcting the quantified microfluidic delivery. **Figure 3** shows the response of a thermal time-of-flight sensor used to detect the bubbles inside the microfluidic channel. (**Figure 3**, Left) The channel design can be found in a previous report [87]. As the two air bubbles (one big and one small) pass through the channel sequentially, the fluidic properties that the sensor senses will be drastically different from those of the pure liquid. In the case of water with air bubbles, the smaller thermal conductivity and substantially lower density for the mixture will prompt a faster heat transfer resulting in a recorded positive flowrate spike. With additional sensing elements to capture the flow speed of the fluid, the time of the burst spike can be used to estimate the sizes of the bubbles. Further, the gas properties inside the bubble might also be detected per the calibration of the sensing element thermal response to the fluid. While in another case, the sensor could also be applied to study the cavitation (**Figure 3**, right). The as-calibrated thermal time-of-flight sensor will normally have an accuracy within ±1%. After calibration in DI water and subsequent verification, the same sensor was kept at the same microfluidic channel at null flow

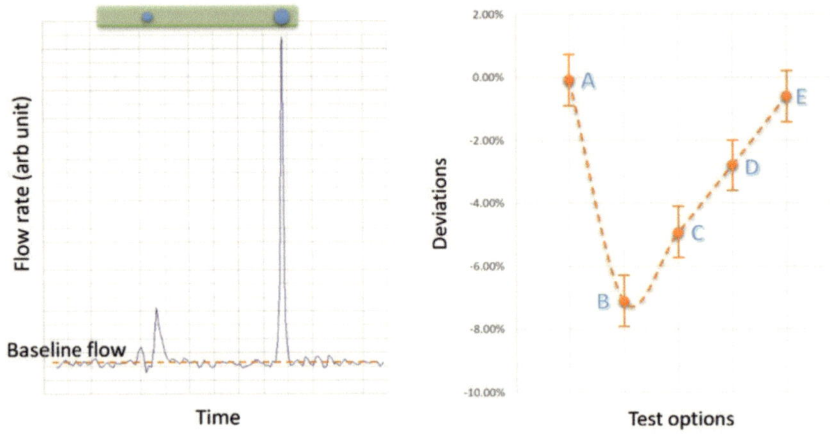

Figure 3.
Left - Example of the response of a micromachined thermal time-of-flight sensor to air bubbles passing in a DI-water microfluidic channel; and right – shows the same sensor response at 20mL/min flow to the channel conditions: A – as calibrated DI water; B – tested after sensor powered on in a null flow DI water channel for 48 hours; C – After B test and degassing for 15 minutes; D – after C and full scale full (30mL/min) flow for 30 minutes; E – after D, the channel dried with N_2 *and re-test.*

condition with a 5 Vdc power applied for 48 hours. The sensor was then re-tested for the registered accuracies by referring to a high precision mechanical syringe pump in serial. One sees that a huge negative deviation of about −7% was recorded (Test B). This could be likely because the sensor's surface had been populated with small air bubbles due to the prolonged constant heat that promoted the bubble nuclei growth and air diffusion. When the flow was started, the drag force might force the collapse of these bubbles causing the cooling that led to the negative deviations. This was further supported by the fact that after degassing the flow microchannel for 15 minutes where the same sensor was installed. After the observed negative deviations, re-measurement of the flow accuracy with the identical procedure, the deviation was reduced (Test C). The deviation was further reduced by running the flow at the full scale for another 30 minutes (Test D). And finally, the sensor recovered to the original precision by dried the sensor surface with nitrogen and the re-test with the same procedure after degassing (Test E), which would effectively eliminate the cavitation by bubbles.

3.3 Microfluidic dissolution

While manipulating the microfluidics inside the designated microchannel, mixing two or more fluids is a common practice. The mixture of the fluids, depending on the physical properties, can be miscible or immiscible. The miscible fluids will result in a fluid with a new *concentration*, while the immiscible fluids will lead to a *Two-phase* fluid. For the gas–liquid mixture, cavitation discussed in the previous section or bubbles will very much likely be formed depending on how dissolvable the gas into the liquid will be. Many studies have been dedicated to the gas–liquid mass transfer, particularly to the Taylor flow-related bubble forming, flowing, and separating, [88] oil-in-water emulsions [89], and other phase-separated immiscible fluids such as carbon dioxide dissolving in various fluids [90]. A gas such as air dissolution in water can decrease nucleation temperature, making the enlargement of the bubble nuclei of water resulting in cooling [91].

Microfluidic dissolution phenomena impose big challenges in metering the flow for a desired metrological accuracy, either with immiscible or miscible fluids.

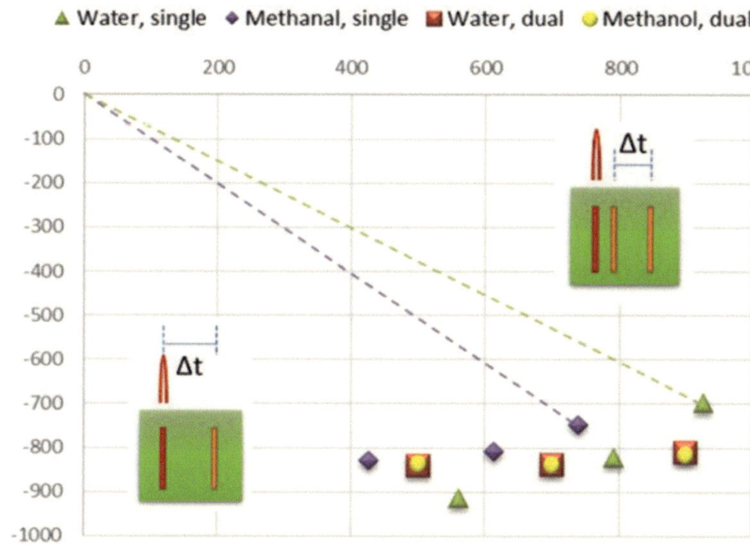

Figure 4.
Thermal time-of-flight measurement of deionized water and methanol flow rates.

The dual-phase or multi-phase flow for the immiscible fluids would involve various liquid–liquid, gas–liquid, liquid–gas–liquid, and supercritical fluid flows beyond the capabilities of the conventional flow sensing approaches. Even with the miscible fluids, the microbubbles would likely present in all cases. The changes in the mixture's density and physical properties will lead to completely different heat and mass transfer, which will significantly deviate the metering values that are always reference to those at the calibration conditions. Optical or image processing would help understand the physical or even chemical process, but it would not help improve the flow measurement accuracy. Therefore, new flow sensing technologies are required for metering these types of microfluidics.

Figure 4 shows the polar plots of a thermal time-of-flight sensor measurement of the deionized water and methanol, respectively, at 3 individual flowrates of 1, 3, and 5 mL/min. The flowrates were set via a precision syringe pump. The sensor's microheater was modulated with a sine wave, and the phase-shifts at the sensing elements were recorded for the flowrate calibration. The fluidic dependent measurement can be seen for the single sensing element configuration as indicated by the differences in measured polar angles between water and methanol. With the dual-sensing elements, the measurements of the two polar plots are overlapped. Therefore, the water calibrated sensor can be directly applied to measure another fluid with different fluidic properties. For the fluidic mixing process with miscible fluids, this dual thermal time-of-flight sensing approach can provide a more desirable measurement than the other thermal sensing approaches. Moreover, as each sensing element's data can be individually acquired, the sensor can also output any changes in its measured fluid. The concentration of the dual miscible fluids can be deduced from the thermal properties measured by comparing the data in the registers at the calibration.

4. Application example: Control of drug infusion

Drug infusion has been in medical practice for over 300 years. Precision control of drug delivery is getting increasing attention in recent years. In a European

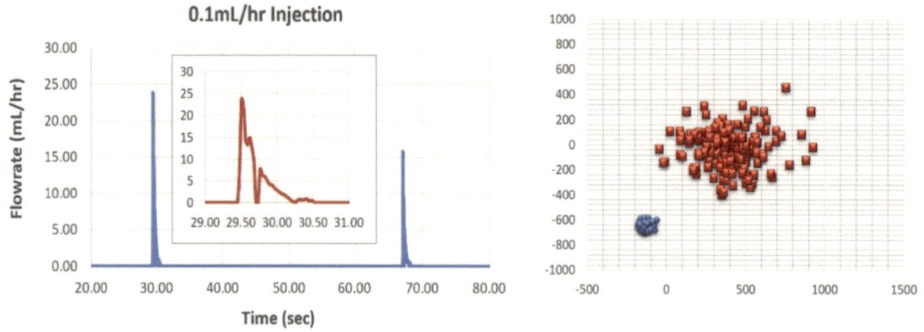

Figure 5.
Drug infusion example: left – commercial infusion pump (Alaris 8100) output at 0.1 mL/hr; and right comparison between the outputs at 20mL/hr by Alaris (red) and a precision syringe pump (blue, KD Scientific Legato 210) measured by a thermal time-of-flight sensor.

Metrology project for drug delivery [92] conducted in 2015, several commercial flow meters with Coriolis, thermal, and differential pressure measurement principles were assessed for metrological performance. However, even the comparisons were made with high precision syringe pumps, some deviations were reported. In practice, many of the devices serving drug infusion are utilizing peristaltic pumps, which have much lower accuracy than the precise syringe pumps [93]. Comparing the peristaltic pump performance and a precise syringe pump can be found in **Figure 5**, the right plot, which is the polar measurements by a thermal time-of-flight sensor at a set point of 20 mL/hr. The red dots are from the peristaltic pump having a large dispersion of the actual flow speed, and the blue dots are from the high precision syringe pump. Therefore, with most of the current drug infusion pumps, the accuracy might not be well controlled since the delivering flow speeds are quite scattered, and a precision sensor is needed to provide the feedback for a close loop. **Figure 5** left plot shows a real-time output of a commercial drug infusion pump Alaris 8100 with a nominal 0.1 mL/hr. delivery speed. From the expanded insert, one sees that the delivery is actually with a pulsed dosage having a wide spectrum of speeds, and the nominal speed is achieved via the adjustment of the time intervals between any of the two pulsed doses. Therefore, the *flowrate* measurement of the flow speeds becomes meaningless, whereas the totalized values would be the ones to provide the real amount of delivered drugs. In an earlier report, [94] a thermal time-of-flight sensor with dual sensing elements suspended in a micromachined microchannel showed a dynamic range of 1000:1 could be achieved. However, to gauge the conventional infusion applications, a sensor with a fast response time of fewer than 3 msec while having a large dynamic range of at least 4000:1 will be needed to meet the requirements for control of total dosage within 5% deviations. A thermal time-of-flight sensor can indeed achieve these conditions with multiple sensing elements.

5. Concluding remarks

Metering the microfluidic flow is critical for many microfluidic applications requiring precise control of the desired microfluidic process or handling. Precision in the flow metering will also improve the performance of the current instrumentation, including the widely applicable drug infusion apparatus, which are nontrivial for the advancement in the medical application and general applications in microfluidics. At the dimensions of interest, current flow sensing technologies are not fully capable of serving the demands. Factors such as fluid and channel interface/interactions, cavitation, and dissolution play critical roles in impacting microfluidic metrology. Additional sensing elements must be integrated with the current flow sensing approaches to compensate, assist, and enhance the flow metrology. In a most recent review, [95] many available technologies can be used to acquire the microfluidic thermodynamic properties such as viscosity, density, diffusion coefficient, solubility, and phase equilibrium directly from the microfluidic channels on a chip. However, many of these technologies are bulky, costly, and not easily integrated with the microfluidic channels. They also often require a transparent microfluidic channel, which would not be readily available in real applications. Although the advancement of micromachining in both the process tooling and application technologies greatly enrich the options for microfluidic flow sensing, a capable device is yet to be demonstrated. The recently developed thermal time-of-flight sensing technologies for microfluidics offer a multiparameter capability and unprecedented dynamic measurement range. The surface acoustic wave flow sensing as a simple yet non-invasive approach is also very promising. Integrating with additional sensing elements and decomposing the acquired information might provide additional viable tools serving to understand, advance, and better control the microfluidic process and handling.

Acknowledgements

The author appreciates his colleagues at Siargao Ltd., who have been dedicated to the challenges and innovations in the commercialization of microfluidic sensing devices since 2009.

Author details

Liji Huang
Siargo Ltd., Santa Clara, California, USA

*Address all correspondence to: liji@Siargo.com

References

[1] Nguyen N T, Wereley S T, Shaegh S A M, Fundamentals and Applications of Microfluidics, 3rd ed. Boston: Artech House; 2019. 576p. ISBN: 9781630813642

[2] Whitesides G: The origin and the future of microfluidics, Nat. 2006; 422:369-373. DOI:10.1038/nature05058

[3] Bassous E, Taub HH, Kuhn L: Inkjet printing nozzle arrays etched in silicon, Appl. Phys. Lett. 1977; 31:135-137. DOI: 10.1063/1.89587

[4] Terry S C, Herman J H, Angell J B: A gas chromatographic air analyzer fabricated on a silicon wafer, IEEE Trans. Electron Devices 1979; 26: 1880-1886. DOI: 10.1109/T-ED.1979.19791

[5] Bhatia SN, Ingber D E: Microfluidic organs-on-chips, Nat. Biotech. 2014; 32: 760-772. DOI: 10.1038/nbt.2989

[6] Convery N, Gadegaard N: 30 years of microfluidics, Micro and Nano Engr. 2019; 2: 76-91. DOI: 10.1016/j.mne.2019.01.003

[7] Manz A, Graber N, Widmer HM: Miniaturized total chemical analysis systems: a novel concept for chemical sensing, Sens. Actuat. 1990; B1: 244-248. DOI: 10.1016/0925-4005(90)80209-I

[8] Shirzadfar H, Khanahmadi M: Review on structure, function, and applications of microfluidic systems, Int. J. Boisen. Bioelectron. 2018; 4: 263-265. DOI: 10.15406/ijbsbe.2018.04.00137

[9] Volpatti L R, Yetisen A K: Commercialization of microfluidic devices. Trends in Biotech. 2014; 32:347-350. DOI: 10.1016/j.tib-tech.2014.04.010

[10] Sackmann E K, Fulton A L, Beebe D J: The present and future role of microfluidics in biomedical research. Nature. 2014; 507 : 181-189. DOI: 10.1038/nature13118

[11] Santos H A, Liu D, Zhang H: editors. Microfluidics for pharmaceutical applications. Amsterdam: Elsevier; 2019. 465p. DOI: 10.1016/c2016-0-03449-6

[12] Burns M A, Johnson BN, Brahmasandra S N, Handique K, Webster J R, et al.: An integrated nanoliter DNA analysis device. Science. 1998; 282: 484-487. DOI: 10.1126/SCIENCE.282.5388.484

[13] Zhang C, Xu J, Ma W, Zheng W: PCR microfluidic devices for DNA amplification. Biotech. Advances. 2006; 24 :243-284. DOI:10.1016/j.biotechadv.2005.10.002

[14] Lei K F: Microfluidic systems for diagnostic applications: a review. J. Lab. Auto. 2012; 17: 330-347. DOI: 10.1177/221106821454853

[15] Mejia-Salazar J R, Cruz K R, Vasques E M M, Oliveira Jr. O N: Microfluidic point-of-care devices: new trends and future prospects for eHealth diagnostics. Sensors. 2020; 1951-1967. DOI: 10.3390/s20071951

[16] Damiani S, Kompella U B, Damiani S A, Kodzius: Microfluidic devices for drug delivery systems and drug screening. Genes. 2018; 9: 103-127. DOI: 10.3390/genes9020103

[17] Ma Q, Cao Y, Han S, Liang Y, Zhang T, et al.: Microfluidic-mediated nano-drug delivery systems: from fundamentals to fabrication for advanced therapeutic applications. Nanoscale. 2020; 12:15512-15527. DOI: 10.1039/D0NR02397C

[18] Ahn J, Ko J, Lee S, Yu J, Kim Y et al.: Microfluidics in nanoparticle

drug delivery; from synthesis to pre-clinic screening. Adv. Drug Del. Rev. 2018; 128:29-53. DOI: 10.1016/j.addr.2018.04.001

[19] Song Y, Cheng D, Zhao L, editors. Microfluidics: Fundamentals, Devices, and Applications: Fundamentals and Applications. Wenham: Wiley; 2018. 567p. DOI: 10.1002/9783527800643

[20] Waheed S, Cabot J M, Macdonald NP, Lewis T, Guijt R M, et al.: 3D printed microfluidic devices: enablers and barriers. Lab on a Chip. 2016; 16:1993-2013. DOI: 10.1039/c6lc00284f

[21] Ghaemi R, Tong J, Gupta BP, Selvaganapathy P R: Microfluidic device for microinjection of *Caenorhabditis elegans*. Micromachines. 2020; 11:295-312. DOI: 10.3390/mi11030295

[22] Postulka N, Striegel A, Krausse M, Mager D, Spiehl D, et al.: Combining wax printing with hot embossing for the design of geometrically well-defined microfluidic papers. ACS Appl. Mater. Interfaces. 2019; 11: 4578-4587. DOI: 10.1021/acsami.8b18133

[23] Blow N: Microfluidics: the great divide. Nat. Methods. 2009; 6:683-686. DOI: 10.1038/nmeth0909-683

[24] Tsao C W: Polymer microfluidics: simple, low-cost fabrication process bridging academic lab research to commercialization production. Micromachines. 2016; 7:225-236. DOI: 10.3390/mi7120225

[25] Ortseifen V, Viefhues M, Wobbe L, Gruenberger A: Microfluidics for biotechnology: bridging gaps to foster microfluidic applications. Front. Bioeng. Biotechnol. 2020; 8: 1-12. DOI: 10.3389/fbioe.2020.589074

[26] Blow N: Microfluidics: in searching of a killer application. Nat. Methods. 2007; 4:665-670. DOI: 10.1038/nmeth0807-665

[27] Yole Developpement, Status of the microfluidic industry 2020. 2020. Available from http://www.yole.fr

[28] Antony R, Nadagopal M SG, Sreekumar N, Selvaraju N: Detection principles and development of microfluidic sensors in the last decade. Microsys. Technol. 2014; 20:1051-1061. DOI: 10.1007/s00542-014-2165-0

[29] Salim A, Lim S: Review of recent metamaterial microfluidic sensors. Sensors. 2018; 18:232-256. DOI: 10.3390/s18010232

[30] Lee G, Lee J, Kim J, Choi HS, Kim J, et al.: Single microfluidic electrochemical sensor system for simultaneous multi-pulmonary hypertension biomarker analysis. Sci. Rep. 2017. 7: 7545-7553. DOI: 10.1038/s41598-017-06144-9

[31] Luka G, Ahmadi A, Najjaran H, Alocilja E, DeRosa M, et al.: Microfluidic integrated biosensors: a leading technology towards lab on a chip and sensing applications. Sensors. 2105; 15: 30011-30031. DOI: 10.3390/s151229783

[32] Klapperich C M: Microfluidic diagnostics: time for industry standards. Exp. Rev. Med. Dev. 2009; 6:211-213. DOI: 10.1586/ERD.09.11

[33] Heeren H: Standards for connecting microfluidic devices? Lab on a Chip. 2012; 12: 1022-1025. DOI: 10.1039/c2lc20937c

[34] Reyes D R, Heeren H: Proceedings of the first workshop on standards for microfluidics. J. Res. Natl. Inst. Stan. 2019; 124: 124001-124022. DOI: 10.6028/jres.124.001

[35] Bissig H, Petter H T, Lucas P, Batista E, Filipe E, et al.: Primary standards for measuring flowrates from

10nl/min to 1 ml/min – gravimetric principle. Biomed. Eng. Biomed. Tech. 2015; 60:301-316. DOI: 10.1515/bmt-2004-0145

[36] Melvin C, Kriihne U, Frederiksen J: Design considerations and initial validation of a liquid microflow calibration setup using parallel operated syringe pumps. Meas. Sci. Technol. 2010; 21: 074004. DOI: 10.1088/0957-02331/7/074004

[37] Beek M P, Lucas P: Realizing primary reference values in the nanoflow regime, a proof of principle. Meas. Sci. Technol. 2010; 21: 074003. DOI: 10.1088/0957-02331/7/074003

[38] Batista E, Godinho I, Martins RF, Mendes R, Robarts J: Development of an experimental setup for microflow measurement using interferometry. Flow Meas. Instr. 2020; 75:101789. DOI: 10.1016/j.flowmeasinst.2020.101789

[39] Zhao P, Cai R, Huang L: A microfluidic flow meter with micromachined thermal sensing elements. Rev. Sci. Instrum. 2020; 91:105006. DOI: 10.1063/1.5131662

[40] Letters F: Liquid flow sensor for nano and microflow ranges. Sensor Review. 2005; 25:20-23. DOI: 10.1108/02602280510577771

[41] Kanne U: Device with flow sensor for handling fluids. 2011; US Patent No.: 7,905,140B2

[42] Yao Y, Chen C, Wu X, Huang L: MEMS thermal time-of-flight meter. In: Proceedings of the 15th International Flow Measurement Conference. 2010; 13-15 October 2010. Taipei, Red Hook: IMEKO; 2011. p. 384-391. ISBN: 9781617824722

[43] High accuracy flow meter for difficult-to-measure microflow of 30mL/min and less Model F7M [Internet]. 2017. Available from: https://www.azbil.com/corporate/technology/techne/techne18_j1.html [Accessed: 2020-12-05]

[44] Jiang H, Wang G: Thermocapacitive Flow Sensor. In: Li D. (eds) Encyclopedia of Microfluidics and Nanofluidics. 2015. Springer: New York. DOI: /10.1007/978-1-4614-5491-5_1566

[45] Kersjes R, Eichholz J, Langerbein Y, Manoli Y, Mokwa W: An integrated sensor for invasive blood-velocity measurement. Sen. Actuat. 1993; A37-38: 674-678. DOI: 10.1016/0924-4247(93)80114-V

[46] International Standard ISO 14511: Measurement of fluid flow in closed conduits – thermal mass flowmeters. 2001; Geneva. www.iso.ch.

[47] Putten A F P, Middelhoek S: Integrated silicon anemometer. Electr. Lett. 1974; 10: 425-426. DOI: 10.1049/el:19740339

[48] Kovasznay L SG: Hot-wire investigation of the wake behind cylinders at low Reynold numbers. Proc. Royal Soc. 1949; 198: 174-190. JSTOR: https://www.jstor.org/stable/98163

[49] Bradbury L J S, Castro I P: A pulsed-wire technique for velocity measurements in highly turbulent flows. J. Fluid Mech. 1971; 49: 657-691. DOI: 10.1017/S0022112071002313

[50] Glatzmaier G C, Ramirez WF: Simultaneous measurement of the thermal conductivity and thermal diffusivity of unconsolidated materials by the transient hot-wire method. Rev. Sci. Instrum. 1985; 56 : 1394-1398. DOI: 10.1063/1.1138491

[51] Yang C, Solberg H: Monolithic flow sensor for measuring milliliter per minute liquid flow. Sen. Actuat. 1992; A33: 143-153. DOI: 10.1016/0924-4247(92)80160-5

[52] van Kuijk J, Lammerink T S J, de Bree H E, Elwenspoek M, Fluitman J H J: Multiparameter detection in fluid flows. Sens. Actuat. 1995; A47: 369-372. DOI: 10.1016/0924-4247(94)00923-6

[53] Huang L, Chen C: Composite MEMS flow sensor on silicon-on-insulator device and method of making the same. 2020; US Patent No. 10,480,974 B2. November 19, 2020.

[54] Bonne U, Kubisiak D, Matthys R J, Schuldt S B: Time lag approach for measuring fluid velocity. 2001; US Patent 6,234,016 B1, May 22, 2001.

[55] Enoksson P, Stemme G, Stemme E: A silicon resonant sensor structure for Coriolis mass-flow measurement. J. Micromech. Syst. 1997; 6: 119-125. DOI: 10.1109/84.585789

[56] Smith R, Sparks D R, Riley D, Najafi N: A MEMS-based Coriolis mass flow sensor for industrial applications. IEEE Trans. Indus. Electr. 2009; 56: 1066-1071. DOI: 10.1109/TIE.2008.926703

[57] Sparreboom W, Geest J, Katerberg M, Postma F, Haneveld J, et al.: Compact mass flow meter based on micro Coriolis flow sensor. Micromachines. 2013; 4:22-33. DOI: 10.3390/mi4010022

[58] Schut T, Wiegerink R, Lotters J: μ-Coriolis mass flow sensor with resistive readout. Micromachines. 2020; 11:184-197. DOI: 10.3390/mi11020184

[59] Fritsche T, Horst-Meyer S Z, Munch H-J: Gas bubble sensing device with two ultrasonic emitters connected to one ultrasonic signal generator. 2017; US Patent No. 9,546,983 B2. Jan. 17, 2017.

[60] Jagannathan H, Yaralioglu G G, Ergun A S, Degertekin F L, Khuri-Yakub B T, et al.: Microfluidic channels with the integrated ultrasonic transducer. 2001 IEEE Proceedings Ultrasonic Symposium. 2: 859-862. DOI: 10.1109/ULTSYM.2001.991856.

[61] Connacher W, Zhang N, Huang A, Mei J, Zhang S et al.: Micro/nano acoustofluidics: materials, phenomena, design, devices, and applications. Lab Chip. 2018; 18:1952-1996. DOI: 10.1039/c8lc00112j

[62] Franke T, Abate A R, Weitz D A, Wixforth A: Surface acoustic wave directed droplet flow in microfluidics for PDMA devices. Lab Chip. 2009; 9: 2625-2627. DOI: 10.1039/b906819h

[63] Wang Y, Li Z, Qin L, Chyu M K, Wang Q M: Theoretical and experimental studies of a surface acoustic wave flow sensor. IEEE Trans. Ultrasonics, Ferreelec. Freq. Ctrl. 2012; 59: 481-490. DOI: 10.1099/TUFFC.2012.2218

[64] Julian F, Azadi S, Razmjou A, Orooji Y, Kottapalli A, et al.: Design and applications of MEMS flow sensors: a review. Sen. Actuat. 2019; A295: 483-502. DOI: 10.1016/j.sna.2019.06.020

[65] Oosterbroek R E, Lammerink T S J, Berenschot J W, Krijnen G J M, Elwenspoek MC, et al.: A micromachined pressure/flow sensor. Sen. Actuat. 1999; 77:167-177. DOI: 10.1016/S0924-4247(99)00188-0

[66] Gass V, van der Schoot B H, de Rooij N F: Nanofluid handling by micro-flow-sensor based on drag force measurement. 1993 Proceedings IEEE Micro Electro Mech. Syst. 1993; P167-172. DOI: 10.1109/MEMSYS.1993.296928

[67] Cheri M S, Latifi H, Sadeghi J, Moghaddam M S, Shahraki H, et al.: Real-time measurement of flow rate in microfluidic devices using a cantilever-based optofluidic sensor. Analyst. 2014; 139:431-438. DOI: 10.1039/c3an01588b

[68] Nezhad A S, Ghanbari M, Agudelo C G, Packirisamy M, Bhat R B, et al.: PDMS microcantilever-based flow sensor integration for lab-on-a-chip. IEEE Sensors J. 2013; 13: 601-609. DOI: 10.1109/JSEN.2012.2223667

[69] North N, Keller S S, Boisen A: Integrated cantilever-based flow sensors with tunable sensitivity for in-line monitoring of flow fluctuations in microfluidic systems. Sensors. 2013; 14: 229-244. DOI:10.3390/s140100229

[70] Zarifi M H, Sadabadi H, Hejazi S H, Daneshmand M, Sanati-Nezhad A: Noncontact and nonintrusive microwave microfluidic flow sensor for energy and biomedical engineering. Sci. Rep. 2018; 8:139. DOI: 10.1038/s41598-017-18621-2

[71] Abdulwahab M R, Ali Y H, Habeeb F J, Borhana A A, Abdelrahman M, et al.: A review in particle image velocimetry techniques. J. Adv. Res. Fluid Mech. Thermal Sci. 2020; 65:213-229. ISSN: 2289-7879

[72] Chung J, Grigoropoulos C P, Greif R: Infrared thermal velocimetry for nonintrusive flow measurement in silicon microfluidic devices. Rev. Sci. Instr. 2003; 74: 2911-2917. DOI: 10.1063/1.1569397

[73] Batista E, Godinho I, Martin RF, Medes R, Robarts J: Development of an experimental setup for a microflow measurement using interferometry. Flow Meas. Instr. 2020; 75:101789. DOI: 10.1016/j.flomeasinst.2020.101789

[74] Stern L, Bakal A, Tzur M, Veinguer M, Mazurski N, et al.: Doppler-based flow rate sensing in microfluidic channels. Sensors. 2014; 14:16799-16807. DOI: 10.3390/s140916799

[75] Lien V, Vollmer F: Microfluidic flow rate detection based on integrated optical fiber cantilever. Lab Chip. 2007; 7:1352-1356. DOI: 10.1039/b706944h

[76] Măriuța D, Colin S, Barrot-Lattes C, Calvé S, Korvink J G, et al.: Miniaturization of fluorescence sensing in optofluidic devices. Microfluid. Nanofluid. 2020; 24:65. DOI: 10.1007/s10404-020-02371-1

[77] Deng S, Wang P, Liu S, Zhao T, Xu S et al.: A novel microfluidic flow rate detection method based on surface plasmon resonance temperature image. Sensors. 2016; 16:964. DOI: 10.3390/s16070964

[78] Collins J, Lee A P: Microfluidic flow transducer based on the measurement of electrical admittance. Lab Chip. 2004. 4:7-10. DOI: 10.1039/b310282c

[79] Alfadhel A, Zaher B L A, Yassine O, Kosel J: A magnetic nanocomposite for biomimetic flow sensing. Lab Chip. 2014; 14: 4362-4369. DOI: 10.1039/C4LC00821A

[80] Acai J, Beebe D: Nature. 2005 ; 437: 648-655. DOI : 10.1038/nature04163

[81] Stone H A, Stroock A D, Ajdari A: Engineering flows in small devices: microfluidics toward a lab-on-a-chip. Annu. Rev. Fluid Mech. 2004; 36: 381-411. DOI: 10.1146/annurev.fluid.36.050802.122124

[82] Wright J D, Schmidt J W: Reproducibility of liquid micro flow measurement. Proceeding of the 18th International Flow Measurement Conference (FLOMEKO2019). 26-28 June 2019. Lisbon. p604-622.

[83] Swann E, Gac S L, Tsuji K, Ohl C-D: Controlled cavitation in microfluidic systems. Phys. Rev. Lett. 2007; 98:254501. DOI: 10.1103/PhysRevLett.98.254501

[84] Ghorbani M, Deprem G, Ozdemir E, Motezakker A R, Villanueva

L G, et al.: On cavitation on-chip in microfluidic devices with surface and sidewall roughness elements. J. Microelectromech. Sys. 2019; 28: 890-899. DOI: 10.1109/JMEMS.2019.2925541

[85] Medrano M, Zermatten P J, Pallone C, Franc J P, Ayala F: Hydrodynamic cavitation in microsystems experiments with deionized water and nanofluids. Phys. Fluids. 2100; 23:127103. DOI: 10.1063/1.3671682

[86] Xu J, Vaillant R, Attinger D: Use of a porous membrane for gas bubble removal in microfluidic channels: physical mechanism and design criteria. Microfluid Nanofluid. 2010; 9:765-772. DOI: 10.1007/s10404-010-0592-5

[87] Luo J, Liu Y, Liao X, Ruan J, Chen C, Huang L: Design of miniature disposable flow sensors for medical applications. In: Proceedings of the 15th International Flow Measurement Conference. 2010; 13-15 October 2010. Taipei, Red Hook: IMEKO; 2011. p. 1016-1021. ISBN: 978-1-61782-472-2

[88] Yang L, Dietrich N, Loubiere K, Gourdon C, Hebrard G: Visualization and characterization of gas-liquid mass transfer around a Taylor bubble right after the formation stage in microreactors. Chem. Engr. Sci. 2016; 143:364-368. DOI: 10.1016/j.ces.2016.01.013

[89] Abedi S, Suteria N S, Chen C C, Vanapalli S A: Microfluidic production of size-tunable hexadecane-in-water emulsions: effects of droplet size on destabilization of two-dimensional emulsions due to partial coalescence. J. Colloid Int. Sci. 2019; 533:59-70. DOI: 10.1016/j.jcis.2018.08.045

[90] Abolhasani M, Singh M, Kumacheva E, Gunther A: Automated microfluidic platform for studies of carbon dioxide dissolution and solubility in physical solvents. Lab Chip. 2012; 12 :1611-1618. DOI : 10.1039/c2lc20143f

[91] Li J, Cheng P: Bubble cavitation in a microchannel. Intl. J. Heat Mass Transfer. 2004; 47:2689-2698. DOI: 10.1016/j.ijheatmasstransfer.2003.11.020

[92] Ogheard F, Batista E, Bissig H, Petter H T, Lucas P, Niemann A K: Metrological assessment of micro-flow meters and drug delivery devices in the scope of the MeDD EMRP project. In: the 17th International Congress of Metrology. 2015; 21-24 September 2015. Paris. EdpSciences: p.09004. DOI: 10.1051/metrology/201509004

[93] Feng Y, Jiang C, Deng W, Yao Y, Chen C, Huang L: Fluid metering for peristaltic pumping. In: Proceedings of the 16th International Flow Measurement Conference. 2013; 24-26 September 2013. Paris. Red Hook: IMEKO; 2014. p. 262-265. ISBN: 978-1-62993-818-9

[94] Berthet H, Jundt J, Durivault J, Mercier B, Angelescu D: Time-of-flight thermal flowrate sensor for lab-on-chip applications. Lab Chip. 2011; 11:215-223. DOI: 10.1039/c0lc00229a

[95] Gavoille T, Pannacci N, Bergeot G, Marlière C, Marre S: Microfluidic approaches for accessing thermophysical properties of fluid systems. React. Chem. Engr. 2019; 4:1721-1739. DOI: 10.1039/c9re00130a

Chapter 5

Micro Milling Process for the Rapid Prototyping of Microfluidic Devices

Muhammad Syafiq Rahim and Abang Annuar Ehsan

Abstract

Micro milling process has become an attractive method for the rapid prototyping of micro devices. The process is based on subtractive manufacturing method in which materials from a sample are removed selectively. A comprehensive review on the fabrication of circular and rectangular cross-section channels of microfluidic devices using micro milling process is provided this review work. Process and machining parameters such as micro-tools selection, spindle speed, depth of cut, feed rate and strategy for process optimization will be reviewed. A case study on the rapid fabrication of a rectangular cross section channel of a microflow cytometer device with 200 um channel width and 50 um channel depth using CNC micro milling process is provided. The experimental work has produced a low surface roughness micro channel of 20 nm in roughness and demonstrated a microflow cytometer device that can produce hydrodynamic focusing with a focusing width of about 60 um.

Keywords: rapid prototyping, micro milling, microflow cytometer, surface roughness, subtractive manufacturing

1. Introduction

The field of microfluidics refers to systems that use millimeter to nanometer-sized fluids for analysis purposes [1]. The system analyzes small samples from micro to nano. Microfluidics is a combination of several fields, such as molecular analysis, molecular biology, biological defense and electronic electronics [2]. Each of these areas contributes to the advancement of microfluidic technology and increased interest in microfluidics [1]. Microflow cytometer are used in the microfluidic flow system, which is a system consisting of a combination of microfluidic and optical, where optical systems are required for analysis purposes [3]. The latest technology for micro flow cytometers focuses on particle focusing to be tested in microfluidics, fluid-controlled shrinkage, optical shrinkage and application integration and integration [3, 4]. Microfluidics are very relevant as they have several advantages, such as requiring only small-sized fluids, and indirectly allowing microfluidics to be tested using micro-to-nano samples [5]. Also, the advantage of microfluidics is that it can be used on small chips. This allows the chip to be used as a portable tool, especially for point of care diagnostic devices (Point of care) (POC). These advantages allow microfluidics to be able to analyze the sample information quickly.

IntechOpen

One of the rapidly evolving phenomena in microfluidic studies is hydrodynamic focusing [6]. Hydrodynamic focusing is a technique that concentrates the flow in the center of the device by manipulating the flow rate of the side [7, 8]. During hydrodynamic focusing the middle path is concentrated by being narrowed by the side path flow. This method of hydrodynamic focusing is important in increasing microfluidic sensitivity [9]. Hydrodynamic focusing can be used to accurately position positions of cells, particles and sensor targets [9]. This hydrodynamic focusing method can be used for the purpose of manipulating cells found in blood composition such as white blood cells and red blood cells [10].

Microfluidic devices can initially be designed using Micro-Electro-Mechanical (MEMS) method [11], the manufacture of silicon-based microfluidic devices usually using this method of Micro-Electro-Mechanical System (MEMS) [3], where its progress is in line with the advancement of semiconductor technology [11]. The processes in this MEMS method involve processes such as oxidation, ion application, low pressure chemical vapor deposition (LPCVD), diffusion, splash, etc. [11].

In addition, for the manufacture of microfluidic rapid prototypes, microfluidic devices can be made using PDMS materials using soft lithographic manufacturing methods or PMMA materials using micro milling.

However, micro milling for microfluidics using PMMA material although a simple and inexpensive method, however, the manufacturing period is longer and not suitable for manufacturing devices in large quantities [12]. This micro milling method for microfluidics is an automated process suitable for the rapid manufacture of prototype devices [12]. Micro milling is a subtractive fermentation process, in which cutting tools are used to remove bulk material from the workpiece. The micro-milling system basically has a work table for XY positions for workpieces, cutting [12].

2. Theory

2.1 Microchannel geometrical shape

The advancement of microfabrication enables the construction of micro channels with micrometer dimensions. Since microfluidic are usually integrated into microsystems, it is important to determine the characteristics of fluid flow in microfluidic for better microfluidic design and operation. From **Figures 1** and **2**, microfluidics can be designed using circular or rectangular shapes. Theoretically, the best form of fluid flow mechanism is a circular channel. But, it is not so noticeable when the device has reached the level of micro or nano scale. First of all, a circular duct has a minimum surface area exposed to fluid that can reduce friction between the wall surface and the liquid. So, the energy required is less to pump water for a given flow

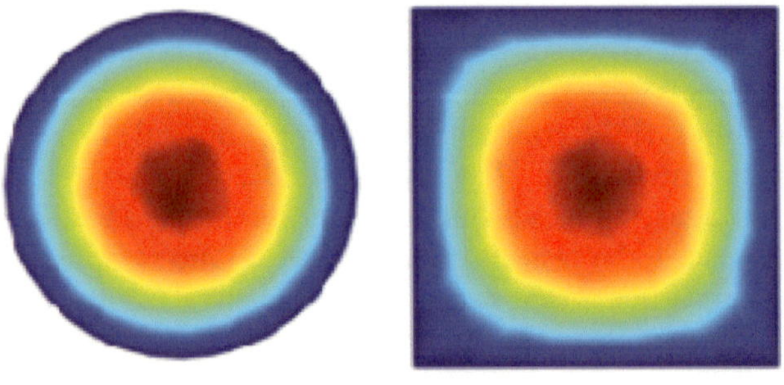

Figure 1.
Microchannel with rectangular and circular geometrical shape [13].

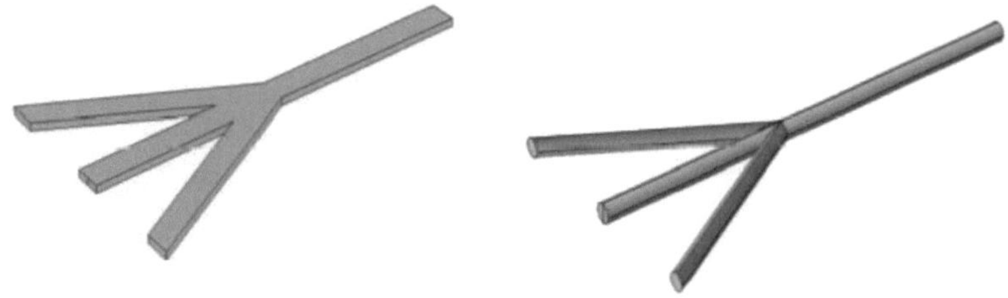

Figure 2.
Microfluidic with rectangular and circular geometrical shape.

rate. Second, the shape of the circle is efficient for handling internal stress. Using a circular channel, the pressure power distribution is uniform across the channel circumference. The presence of sharp corners in the rectangular duct will focus on the edges and sometimes this area needs to be strengthened to resist pressure.

Cell traps with hydrodynamic methods also show the advantage of a round shape to isolate a cell by reducing the applied pressure. By doing this, cells will have a higher percentage to survive in extreme flow conditions [14]. The purpose of this study was to simulate the flow of fluid in the micro-channel using COMSOL Microfluidic. The rectangle was chosen because it is widely used during the fabrication process of PDMS devices with soft lithography.

2.2 Surface roughnes of microchannel

The terms surface roughness and surface finish are widely used in the manufacturing sector to measure surface after machining. Average roughness is the arithmetic mean of the surface roughness profile measure of the mean line, and is the most widely used and universally recognized surface roughness parameter. The surface roughness of the machine in the final micro milling process depends on commonly used process parameters such as tool geometry, spindle speed, feed rate and depth of cut [15]. There are other factors of the micro milling process that affect the surface roughness such as the tip of the micro milling, the breakdown of the tool, the breakdown of the tool (and the nature of the workpiece which has a high quality surface).

Therefore, factors such as vibration and chip removal where these factors are not critical in the macro scale, can have a significant impact on the surface produced on the micro scale. The surface produced after micro milling is found to be affected by the end radius of the micro-tool and the feed rate. It is reported that when the 2 μm of the end radius, and in the state of the feed rate is reduced, the surface roughness increases, indicating that, the optimal feed rate can produce the lowest surface roughness. Cutting speed d an cutting depth affects the surface roughness on the PMMA material [16]. Further, it is found from previous studies as well, depth cutting has the greatest impact while, cutting speed has the lowest effect [16]. Surface roughness also depends on machining parameters and workpiece conditions, tool and heat conditions were also found to affect surface roughness [16]. In addition, the resulting surface quality after machining can be improved by increasing the rigidity and accuracy of the equipment. Because there are various manufacturing methods for polymer-based microfluidics, changes in the surface of the polymer after the manufacture of microfluids attract the interest of many researchers. Many researchers have tried various methods to reduce surface roughness for microfluidics to improve optical quality and improve biological capabilities.

Table 1 shows the surface roughness produced using the micro milling technique. Based on previous studies, it was found that the surface roughness produced by the micro milling can reach up to as little as 38 nm. However, surface

Diameter	Material	Spindle speed	Feed rate	Depth of cut	Surface roughness	Reference
0.8 mm	Carbide	2000 rpm	2 mm/min	1.5 μm	0.352 μm	[17]
0.45 mm	Diamond Coated	150,000 rpm	5 μm/flute	50 μm	38 nm	[18]
0.2 mm	N/A	20,000 rpm	300 mm/min	10 μm	0.13 μm	[19]
0.1 mm to 0.5 mm	Carbide	10,000 rpm	20 mm/min	10–20 μm	70–85 nm	[20]
0.8 mm	Carbide	30,000 rpm	2.65 mm/min	40 μm	128.24 nm	[21]

Table 1.
Surface roughness using different of material, spindle speed, feed rate and depth of cut.

roughness can be achieved up to 38 nm if the micro tool used is coated with the diamond. Micro–tool coated with high-cost diamond are not an option for micro-manufacturing.

3. Case study

3.1 Design of microfluidic

Since this study uses a micro milling a microfluidic design with a rectangular geometry will be used. From **Figure 3**, the designed depth is 50 um, 200 um wide, and the circle on the inlet and outlet has a diameter of 0.6 um. **Figure 3** shows microfluidics with 2 layer PMMA to be fabricated. From **Figure 3**, the top layer has 4 holes with a diameter of 0.8 mm, the design of the hole is based on the need to place a tube with an outer diameter of 0.7 mm. While the design for the bottom layer of microfluidics, there is a circular inlet and outlet with a diameter of 0.6 mm which is smaller than the outer diameter of the tube, to allow the tube to be above the microfluidic layer and the entire fluid can enter the micro flow.

The tool used in this research is a 0.2 mm diameter tool made of carbide material, has 2 flutes and Aluminum coated. While the workpiece that will be used in this research is Poly (methyl methacrylate) or referred to as acrylic which has a thickness of 2 mm.

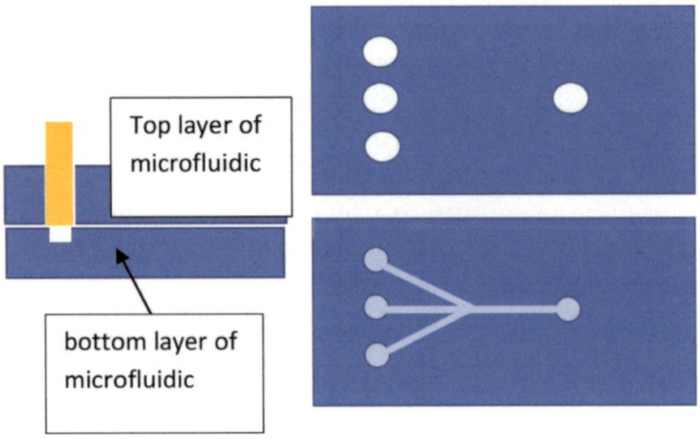

Figure 3.
Top and bottom layer of microfluidic.

3.2 Fabrication of microfluidic using micromilling

The Taguchi method as shown in **Table 2** is used taking into account 3 main parameters, namely, spindle speed, cutting depth, and feed rate to obtain the lowest surface roughness. The Taguchi method which uses 3 parameters along with 3 stages is used as **Table 2**. Spindle speeds consisting of 4000 rpm, 5000 rpm, and 6000 rpm, and spindle speeds lower than 10,000 rpm are used because PMMA material will burn when high speeds are used, as high speeds can increase the temperature on the tool can cause micro flow size the result is greater than desired. The cutting depths used for each cut are 0.01 μm, 0.025 μm, and 0.05. This is to ensure that the discarded chip is smaller than the tip of the tool. While the feed rate used is 10 mm/min, 15 mm/min, and 20 mm/min. Due to the high feed rate it can cause the tool to break. The total number of experiments produced is 9 experiments as shown in **Table 3**, each surface roughness average will be recorded, based on the smallerer the better method, and the smallest surface roughness average parameter will be taken. Then the optimal parameters will be repeated 10 times to ensure that the parameters produce consistent and stable results.

After analyzing the experimental data from **Table 4**, the lowest surface roughness can be obtained by using a spindle speed of 4000 rpm, a feed rate of 10 mm/min and a depth cut of 0.01 mm. However, based on **Table 4**, it can be seen that while the spindle speed is 6000 rpm, cutting depth and feed rate do not have a significant impact on surface roughness, where the average surface roughness is recorded around 100 nm to 200 nm, at the same time, increasing cutting depth and feed rate, increasing average surface roughness resulting. Moreover, it can be observed that all the resulting surface roughness is less than 450 nm. Next, to validate the experiment, 10 microcontrollers were built on PMMA with spindle speed parameters of 4000 rpm, feed rate of 10 mm/min and depth depth of 0.01.

Factors	Level 1	Level 2	Aras 3
Spindle speed (rpm)	4000	5000	6000
Depth of cut (μm)	0.01	0.025	0.05
Feed rate (mm/min)	10	15	20

Table 2.
Machining parameter (Taguchi method).

Experiment number	Spindle speed (rpm)	Depth of cut (mm)	Feed rate (mm/min)
1	4000	0.010	10
2	4000	0.025	15
3	4000	0.050	20
4	5000	0.010	15
5	5000	0.025	20
6	5000	0.050	10
7	6000	0.010	20
8	6000	0.250	10
9	6000	0.050	15

Table 3.
Experiment number (Taguchi method).

Number	Spindle speed (rpm)	Depth of cut (mm)	Feed rate (mm/min)	Surface roughness (nm)
1	4000	0.01	10	67.3018
2	4000	0.025	15	267.2102
3	4000	0.05	20	406.8926
4	5000	0.01	15	170.2524
5	5000	0.025	20	350.468
6	5000	0.05	10	442.6494
7	6000	0.01	20	119.4901
8	6000	0.025	10	139.6821
9	6000	0.05	15	170.2192

Table 4.
Surface roughness by using different machining parameters.

During the machining process, a drop of water is placed on the substrate to remove debris during machining. The average surface roughness obtained from 10 validation experiments is shown in **Table 5**, where the average roughness is 24.0824 nm with a standard deviation of 4.2509 nm.

Selecting the cutting depth range and the feed rate with less than the minimum value will result in an increase in machining time, however, the cutting depth value, spindle speed and high feed rate, can increase the risk of damaged tool as reported [22]. From **Table 4** a total of 9 microchannel with a depth of 50 μm and a width of 200 μm were tested using the Alicona Infinite Focus Microscopy (IFM) 3D Optical Profiler used to measure the roughness of the surface on the cut of microchannels. The area of surface roughness shown at **Figure 4**. Analytical factors can be used to determine the main cutting parameters in the micro milling of the PMMA substrate. Based on **Table 4**, the larger the resulting range, the greater the influence of these factors on surface roughness, in this research, the depth of cutting indicates the largest range. This shows that the depth of cutting has a great influence on surface roughness. Whereas, the feed rate indicates a low range, this means that the feed rate has the least influence on surface roughness.

Number	Surface roughness (nm)
1	21.3106
2	20.1148
3	26.7489
4	23.628
5	19.3741
6	23.5145
7	22.9668
8	27.5627
9	33.6486
10	21.9548
Average	24.08238

Table 5.
Surface roughness by using optimal machining parameters.

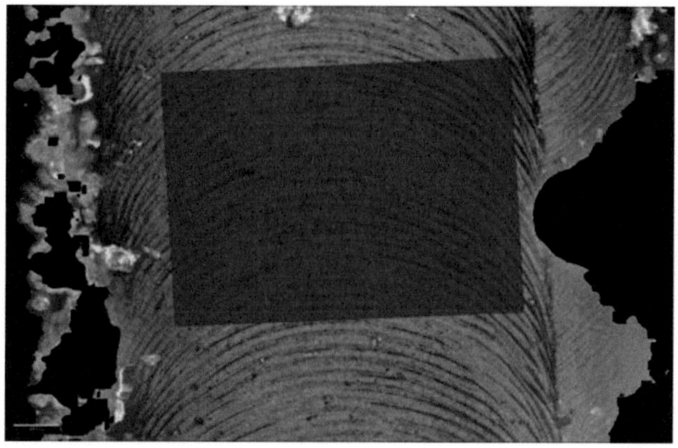

Figure 4.
Area for surface roughness measurement using infinite focus microscopy.

Table 4 also shows the optimal cutting parameters for obtaining minimal surface roughness. **Table 4** shows the combination of machining parameters to obtain the smallest surface roughness is the spindle speed 4000 rpm, cutting depth 0.01 μm and feed rate 10 mm/min. Based on **Table 4**, the average surface roughness average achieved for this parameter is 67.3018 nm. Moreover, from this study, based on **Figure 5**, if the study is compared by looking at the same parameter readings, shows that the spindle speed of 6000 rpm can produce the lowest surface roughness compared to the spindle speed of 4000 rpm and 5000 rpm. It shows that the cutting depth of 0.01 μm produces the lowest surface roughness followed by 0.025 mm and 0.05 mm. Furthemore, the feed rate of 15 mm/min produces the lowest surface roughness followed by 10 mm/min and 20 mm/min. Based on **Table 4**, it shows that the cutting depth most influences the roughness of the resulting surface followed by spindle speed and feed rate. This is in line with the theory that low cutting depths can result in low chip loads, this allows lower surface roughness to be achieved. As previously discussed, low depth of cut can result in low surface roughness.

3.3 Hydrodynamic focusing experiment

After successful microfluidic installation, the experiment was continued by testing the hydrodynamic focus. This feature is important to ensure that the designed microfluidics can operate, there are several factors that can cause the microfluidics to be unable to operate, firstly due to clogged microwaves, secondly because the bond between the 2 wafers is not strong causing small holes that cause leakage. Based on **Table 6**, the resulting focusing width is related to the sheath and sample flow rate ratio. The resulting focusing width can be adjusted according to the desired application. However, the sample flow width must be adjusted according to the specific cell size for detection, at the same time, allowing cells to pass through them one by one on the sample flow, this is to increase the sensitivity of the constructed device. Reynold numbers are kept in low condition, this is to avoid uninterrupted flow of microfluidics [23].

Based on this hydrodynamic focusing experiment shown at **Figure 6**, the side path with a flow rate of 3000 μl/min and the flow rate for the sample path of 10 μl/min can produce a focusing width as low as 39 μm. However, with an sheath flow rate of 3000 μl/min and a sample path flow rate of 100 μl/min, the resulting focusing width is 60 μm. Both of these results answer for the objective of the study, namely the production of hydrodynamic focusing around 60 μm. Based on **Table 6** it can also be observed, that if a flow rate ratio of 10 and 100 is used, a

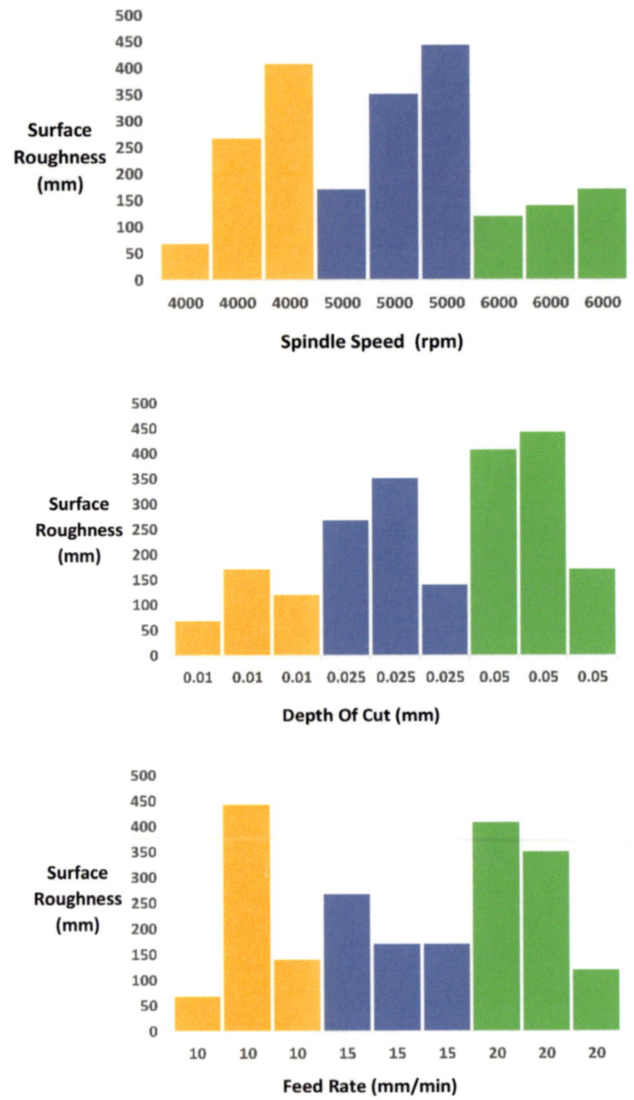

Figure 5.
Result of surface roughness by comparing same machining parameter.

Ratio	Sheath flow rate	Sample flow rate	Width of hydrodynamic focusing
10	1000 μl/min	100 μl/min	89 μm
100	1000 μl/min	10 μl/min	67 μm
30	3000 μl/min	100 μl/min	60 μm
300	3000 μl/min	10 μl/min	39 μm

Table 6.
Width of hydrodynamic focusing.

focusing width of 67 μm and 89 μm can be produced. From the simulation results show that effective hydrodynamic focusing occurs only when the sheath flow rate is higher than the center flow rate.

Furthermore, from the simulation results of nonlinear behavior will occur when too high a ratio is used. Increasing the ratio of sheath flow rate to large central

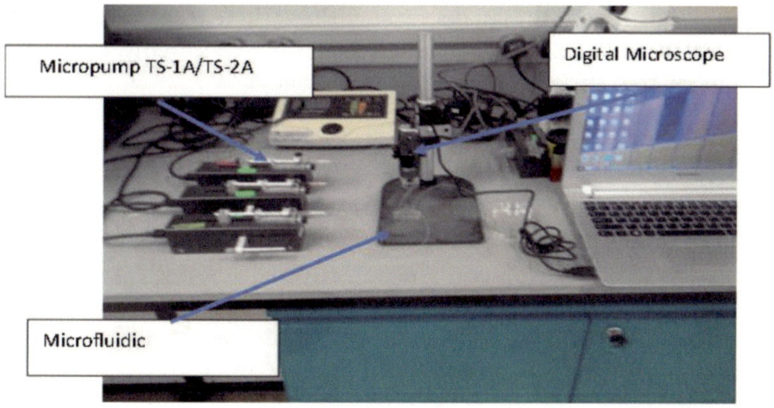

Figure 6.
Setup for hydrodynamic focusing experiment.

flow will only have a small effect on hydrodynamic focusing and may even cause hydrodynamic focusing not to occur if too high a flow rate is used. As shown in **Figure 7**, the hydrodynamic focusing that occurs is in the state of laminate and fully developed. This experiment can also give the impression that the bonding technique between 2 PMMA wafers using ethanol material was successfully performed, since hydrodynamic focusing can be formed. However, it should be noted that the hydrodynamic focus that occurs is not only due to the inflow rate by the fluid only, but the microfluidic geometry constructed also has a significant impact on the characteristics of hydrodynamic focusing. Especially when taking into account the rectangular geometry is easier to do by a micro milling than a round design. The forces formed to control hydrodynamic focusing are more complex than hydrodynamic focusing calculated only on the flow rate ratio [24].

An important aspect of designing and operating for the purpose of hydrodynamic focusing is to identify the position of the focus flow formed. Both lateral flows should have the same flow rate to ensure that the focusing flow flows in the middle of the micro flow. If the asymmetric focusing flow, the focusing flow will be deducted from the flow axis. Based on **Figure 7**, it can be observed that the

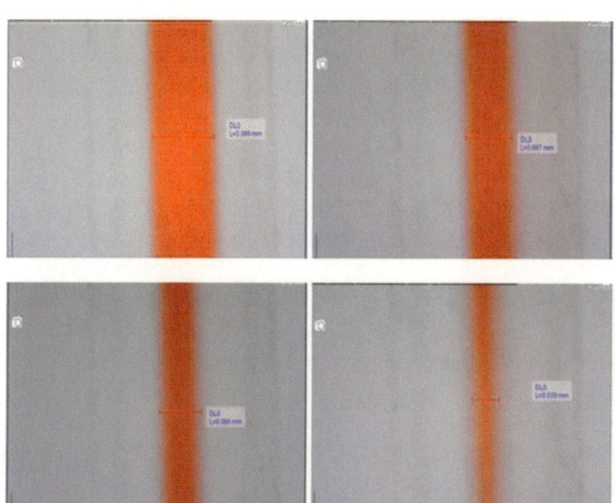

Figure 7.
Hydrodynamic focusing.

hydrodynamic focusing width will decrease if the ratio (sheath flow rate to the main flow rate) increases. It can also be observed from the experiment, it shows that the results of focusing will shrink if a larger ratio is used.

4. Conclusion

In this study, simple, low-cost and real-time methods have been used to detect touch between tools and workpieces. The objective of this study is to find the optimal parameters to achieve low surface roughness using a micro milling, from the data trends obtained, the parameters to obtain the lowest surface roughness are 4000 rpm spindle speed, 10 mm/min billing rate and 0.01 μm cutting depth. However, from the data obtained as well, it shows that water droplets placed on the tool during cutting also contribute to the reduction of surface roughness. In addition, there are several other parameters that can be studied in the future, namely the tool material (e.g. diamond), the smaller tool size, and the type of coolant used.

Since microfluidics manufacturing studies designed using micro milling are still limited, its function can be tested by looking at the hydrodynamic focusing that occurs. This study uses 2 PMMA-based wafers and is bonded using thermal-assisted ethanol. Based on the experiments conducted, the resulting hydrodynamic focusing has a width as small as 39 μm if the sheath flow rate and the center flow rate used are 3000 μl/min and 100 μl/min. Apart from using fluids such as water and dyes, fluids that have properties such as blood can also be used so that more accurate results can be produced.

Acknowledgements

The authors would like to thank Universiti Kebangsaan Malaysia, in particular the Nanophotonics Lab, Institute of Microengineering and Nanoelectronics (IMEN) for providing the facilities and for supporting this research through the Research University Grant scheme, GUP-2019-071.

Author details

Muhammad Syafiq Rahim and Abang Annuar Ehsan*
Institute of Microengineering and Nanoelectronics (IMEN), Universiti Kebangsaan Malaysia (UKM), Bangi, Malaysia

*Address all correspondence to: aaehsan@ukm.edu.my

References

[1] E. K. Sackmann, A. L. Fulton, and D. J. Beebe, "The present and future role of microfluidics in biomedical research.," Nature, vol. 507, no. 7491, pp. 181-189, Mar. 2014, doi: 10.1038/nature13118.

[2] G. M. Whitesides, "The origins and the future of microfluidics.," Nature, vol. 442, no. 7101, pp. 368-373, Jul. 2006, doi: 10.1038/nature05058.

[3] R. J. Yang, L. M. Fu, and H. H. Hou, "Review and perspectives on microfluidic flow cytometers," Sensors Actuators, B Chem., vol. 266, pp. 26-45, 2018, doi: 10.1016/j.snb.2018.03.091.

[4] D. A. Ateya, J. S. Erickson, P. B. Howell, L. R. Hilliard, J. P. Golden, and F. S. Ligler, "The good, the bad, and the tiny: A review of microflow cytometry," *Analytical and Bioanalytical Chemistry*, vol. 391. pp. 1485-1498, 2008, doi: 10.1007/s00216-007-1827-5.

[5] B. K. Gale *et al.*, "A review of current methods in microfluidic device fabrication and future commercialization prospects," *Inventions*, vol. 3, no. 3, 2018, doi: 10.3390/inventions3030060.

[6] M. A. Daniele, D. A. Boyd, D. R. Mott, and F. S. Ligler, "3D hydrodynamic focusing microfluidics for emerging sensing technologies," Biosens. Bioelectron., vol. 67, pp. 25-34, 2015, doi: 10.1016/j.bios.2014.07.002.

[7] M. Lu *et al.*, "Microfluidic hydrodynamic focusing for synthesis of nanomaterials," *Nano Today*. 2016, doi: 10.1016/j.nantod.2016.10.006.

[8] M. Dziubinski, "Hydrodynamic Focusing in Microfluidic Devices," in *Advances in Microfluidics*, 2012.

[9] J. P. Golden, G. A. Justin, M. Nasir, and F. S. Ligler, "Hydrodynamic focusing-a versatile tool," *Anal. Bioanal. Chem.*, 2012, doi: 10.1007/s00216-011-5415-3.

[10] A. Rajawat and S. Tripathi, "Disease diagnostics using hydrodynamic flow focusing in microfluidic devices: Beyond flow cytometry," *Biomed. Eng. Lett.*, no. 0123456789, 2020, doi: 10.1007/s13534-019-00144-6.

[11] F. J. Galindo-Rosales, "Complex fluid-flows in microfluidics," Complex Fluid-Flows Microfluid., pp. 1-111, 2017, doi: 10.1007/978-3-319-59593-1.

[12] D. J. Guckenberger, T. de Groot, A. M.-D. Wan, D. Beebe, and E. Young, "Micromilling: A method for ultra-rapid prototyping of plastic microfluidic devices," *Lab Chip*, vol. 15, no. 11, 2015, doi: 10.1039/C5LC00234F.

[13] T. Nguyen and W. Park, "Rapid, Low Cost Fabrication Of Circular Cross-Section Microchannels By Thermal Air Molding," pp. 348-350, 2015.

[14] M. F. M. R. Wee, M. R. Buyong, and B. Y. Majlis, "Effect of microchannel geometry in fluid flow for PDMS based device," *RSM 2013 IEEE Reg. Symp. Micro Nanoelectron.*, pp. 391-393, Sep. 2013, doi: 10.1109/RSM.2013.6706573.

[15] P. Maurya, P. Sharma, and B. Diwaker, "Implementation of Taguchi methodology to Optimization of CNC end milling process parameters of AL6351 – T6," vol. 2, pp. 3530-3533, 2012.

[16] P.-C. Chen, C.-W. Pan, W.-C. Lee, and K.-M. Li, "An experimental study of micromilling parameters to manufacture microchannels on a PMMA substrate," Int. J. Adv. Manuf. Technol., vol. 71, no. 9-12, pp. 1623-1630, Jan. 2014, doi: 10.1007/s00170-013-5555-z.

[17] M. Y. Ali, A. R. Mohamed, B. Asfana, M. Lutfi, and M. I. Fahmi, "Investigation of Vibration and Surface Roughness in Micro Milling of PMMA," *Appl. Mech. Mater.*, vol. 217-219, no. November, pp. 2187-2193,

2012, doi: 10.4028/www.scientific.net/amm.217-219.2187.

[18] E. Korkmaz, R. Onler, and O. B. Ozdoganlar, "Micromilling of Poly(methyl methacrylate, PMMA) Using Single-Crystal Diamond Tools," Procedia Manuf., vol. 10, pp. 683-693, 2017, doi: 10.1016/j.promfg.2017.07.017.

[19] P. C. Chen, Y. C. Chen, C. W. Pan, and K. M. Li, "Parameter optimization of micromilling brass mold inserts for microchannels with Taguchi method," Int. J. Precis. Eng. Manuf., vol. 16, no. 4, pp. 647-651, 2015, doi: 10.1007/s12541-015-0086-1.

[20] A. Aramcharoen, S. K. C. Sean, and L. Kui, "An experimental study of micromilling of polymer materials for microfluidic applications," *Int. J. Abras. Technol.*, vol. 5, no. 4, p. 286, 2013, doi: 10.1504/ijat.2012.052037.

[21] N. Atiqah, M. Y. Ali, A. R. Mohamed, and M. S. H. Chowdhury, "Investigation of Surface Roughness and Material Removal Rate for High Speed Micro End Milling on PMMA," *Adv. Mater. Res.*, vol. 1115, no. November 2016, pp. 12-15, 2015, doi: 10.4028/www.scientific.net/amr.1115.12.

[22] E. Kuram and B. Ozcelik, "Optimization of machining parameters during micro-milling of Ti6Al4V titanium alloy and Inconel 718 materials using Taguchi method," *Proc. Inst. Mech. Eng. Part B J. Eng. Manuf.*, 2015, doi: 10.1177/0954405415572662.

[23] M. Nasir, D. R. Mott, M. J. Kennedy, J. P. Golden, and F. S. Ligler, "Parameters affecting the shape of a hydrodynamically focused stream," pp. 119-128, 2011, doi: 10.1007/s10404-011-0778-5.

[24] S. Torino, "Engineering of Microfluidic Devices For Cell Hydrodynamic Manipulation (Thesis)," p. 124, 2015.

Chapter 6

Micromixers for Wastewater Treatment and Their Life Cycle Assessment (LCA)

Olga P. Fuentes, Mabel J. Noguera, Paula A. Peñaranda, Sergio L. Flores, Juan C. Cruz and Johann F. Osma

Abstract

The use of micromixers and catalytically active nanocomposites can be an attractive alternative for the treatment of wastewaters from the textile industry, due to their high activity, low consumption of such nanocomposites, short reaction times and the possibility to work under continuous operation. In this study, 6 different designs of micromixers were modeled and evaluated for the treatment of wastewaters. Velocity profiles, pressure drops, and flows were analyzed and compared for the different devices under the same mixing conditions. In addition, Life cycle assessment (LCA) methodology was applied to determine their performance in terms of environmental impact. Considering the high environmental impact of water sources contaminated by dyes from the textile industry, it becomes critically important to determine when the proposed micromixers are a suitable alternative for their remediation. The LCA and operational efficiency studies results shown here provide a route for the design of novel wastewater treatment systems by coupling low-cost and high-performance micromixers.

Keywords: micromixers, wastewater, life cycle assessment, dyes, magnetite

1. Introduction

Microfluidics is the science that study fluid behavior on micro/nano scales that are circulating in artificial microsystems [1, 2]. Also, this science considers the fabrication of fluidic devices for the transport, delivery, and handling of fluids on the order of microliters or even smaller volumes [3]. Microfluidic techniques have shown advantages such as high performance, design flexibility, low reagent consumption, miniaturization, and automation [4]. The application of these techniques has led to microfluidic devices that have found application in several fields, such as medical and biochemical analysis, environmental monitoring, biochemical, and microchemistry [5–8].

Currently, there is a growing need to monitor water quality across a broad range of applications, including industrial wastewaters as well as drinking water and different surface waters (rivers, lakes, groundwater and marine) [9]. Water sources contaminated by dyes or phenolic compounds, which are present in textile industrial wastewater, represent a threat to human health and the environment [10]. For that reason, it is imperative to find efficient routes to monitor these pollutants in wastewater in order to avoid their discharge above permissible limits. A wide range of sensors and analyzers are commercially available for wastewater monitoring, and they are based on different detection techniques, such as colorimetric, chemical, electrochemical or optical [11]. Here, a growing trend is emerging where microfluidic technologies are considered for environmental detection mainly due to their lower investment and operation costs, as well as reduced infrastructure requirements. Moreover, it has been shown that microreactors, help to maximize biodegradation processes due to the absence of dead volume, allow to perform continuous reactions, and enable to control the contact between the reagents by changes in the microchannel geometries [9, 12].

By handling fluids in microchannels, it is possible to achieve high production yields, and minimize waste generation. Moreover, with this approach it is feasible to operate under short reaction and analysis times, is relatively cheap and enable high-throughput schemes [13]. However, the manufacture of microfluidic devices generally relies on sophisticated cleanroom techniques [14], which is disadvantageous due to their high costs. This issue has been overcome with low-cost manufacturing methods such as polymer laminates, 3D printing, and laser cutting [15, 16]. In this approach, devices are often manufactured by cutting a piece of polymethylmethacrylate (PMMA) followed by engraving a predesigned microchannel pattern on a separate PMMA. The device is then assembled by gluing the two pieces together. PMMA is one of the preferred thermoplastics for the manufacture of microfluidic devices, due to its optical transparency, superior mechanical properties, low cost and good workability in conjunction with its ease for prototyping and mass manufacturing [17]. In this study, we will how PMMA can be used to manufacture micromixers and we will analyze and compare the potential environmental impact of implementing them for wastewater treatment.

Life cycle assessment (LCA) has been widely applied in the wastewater treatment industry due to its important role as a tool for the sustainability assessment of new technologies, processes and the improvement of waste management practices. On this, inputs, such as raw materials and energy, and outputs, such as waste and emissions, are collected in the form of elementary flows for the whole life cycle (Life Cycle Inventory – LCI step) and then converted into environmental impact indexes by means of characterization factors (Life Cycle Impact Assessment – LCIA step) [18]. According to Corominas et al. [19], LCA can be a useful decision-support tool for examining alternative future operational scenarios during strategic planning within the water sector. Also, LCA evaluates beyond the limit imposed by the trade-off between process efficiency and final effluent quality because it considers resource and energy consumption, air emissions and waste generation [20].

In this study, we explore the design and manufacture of micromixers for wastewater treatment to enable the enzyme-based degradation of dyes. In this regard, we propose a LCA assessment to establish the potential environmental impact of implementing these devices. Also, this analysis integrates the required chemical supplies, energy, and water needed for wastewater treatment. Through life cycle assessment (LCA), we compared six different designs of micromixers to identify the one providing the least environmental impact during operation. LCA analysis might therefore contribute significantly to improving wastewater treatment process by coupling micromixers capable of remediating wastewaters with high efficiencies.

2. Materials and methods

2.1 Materials

Iron (II) chloride tetrahydrate (98%) ($FeCl_2$*$4H_2O$), Iron (III) chloride hexahydrate (97%) ($FeCl_3$*$6H_2O$), and dye Eriochrome Black T (EBt) (C.I. 14645) were obtained from PanReac AppliChem (Spain). 2,2-azino-bis(3-ethylbenzothiazoline-6) sulphonic acid (ABTS), glutaraldehyde (25%), sodium hydroxide (NaOH) (98%), tetramethylammonium hydroxide (TMAH) (25%), 3-Aminopropyl-triethoxysilane (APTES) (98%) were purchased from Sigma-Aldrich (USA). Polymethyl methacrylate (PMMA), Methyl methacrylate, Ethanol (96%) and 345 mT Neodymium cylindrical magnets (ϕ: 6 mm x h: 7 mm) were purchased at a local shop.

2.2 Laccase

Laccases (*P. sanguineus* CS43) (EC 1.10.3.2) were obtained from tomato medium as described elsewhere [21]. Briefly, mycelia were removed from the culture supernatant by filtration using two tangential flow filters in series, one of them with pore size of 0.5 mm while the other of 0.2 mm. The obtained laccase cocktail was ultra-filtered using a membrane with a molecular weight cut-off of 10 kDa.

2.3 Synthesis of magnetite and laccase immobilization

Magnetite nanoparticles were synthesized by coprecipitation of 20 mL of 1 M $FeCl_2$ and 20 mL of 2 M $FeCl_3$ under agitation at 1,500 rpm and 90 °C. 40 mL of 8 M NaOH and 40 mL of 2% (v/v) TMAH were then added to the mixture during 3.5 h at a flow rate of 12 mL/h. Nanoparticles (Magnetite) were magnetically separated aided by a strong permanent magnet, then washed thoroughly with 2% (v/v) TMAH, and finally sonicated for 100 min in a VibraCell ultrasonic bath (Sonics, USA).

Magnetite nanoparticles were buffered by adding a NaOH solution until pH approached 11, then sonicated for 10 min. 50 μL of 2% (v/v) TMAH was added and dispersed and then the mixture was sonicated for 10 min. Silanization of the nanoparticles was carried out by adding 50 μL of 2% (v/v) APTES followed by sonication for 20 more min. 50 μL of 2% (v/v) glutaraldehyde was added to the mixture as the crosslinker, and left to react for 30 min. Finally, 50 μL of 960 U/L laccase was added and left overnight to immobilize the enzyme on the nanoparticles. The resulting bionanocompounds (i.e., Lac-Magnetite) were separated by magnetism and washed thoroughly with MilliQ water.

2.4 Geometry design and fabrication

Six different prototypes of micromixers were designed with different microchannel geometries for the reaction chambers. This was achieved by varying the number of layers of PMMA sheets required to create the channel. In the case of one layer, the channel geometry was circular and triangular. In contrast the assembly of multiple layers enabled rectangular-3D, one loop, two horizontal loops, and two vertical loops (See **Figure 1**).

For the micromixers manufacture, each design was engraved and cut on sheets of polymethylmethacrylate (PMMA), with a thickness of 3 mm and an area of 75x25 mm, using a Speedy 100, 60 W laser cutting system (TROTEC, Germany). Sheets were glued together to assemble the devices by applying a few drops of 96% ethanol on the contacting surfaces and maintaining a constant pressure for 8 minutes at 105 °C.

2.5 Experimental test for wastewater treatment

To estimate the dye biodegradation, we selected the EBt dye as a model. The EBt solutions were prepared at pH 5.48 and three different concentrations, namely, 5 mg/L, 10 mg/L and 20 mg/L. Biodegradation tests were conducted by introducing 5 mg of the bionanocompound and 5 mL of dye solution into each micromixer for 25 minutes at a constant rate of 12 mL/h. A neodymium permanent magnet, of

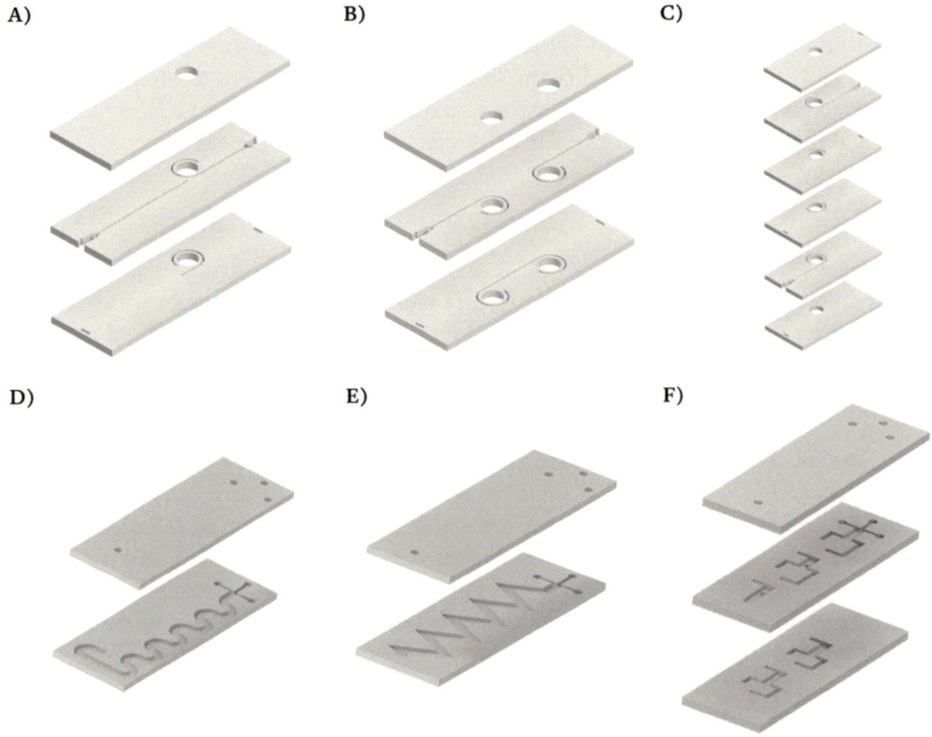

Figure 1.
Micromixers geometries. A) One loop, B) two horizontal loops, C) two vertical loops, D) circular, E) triangular and F) rectangular-3D.

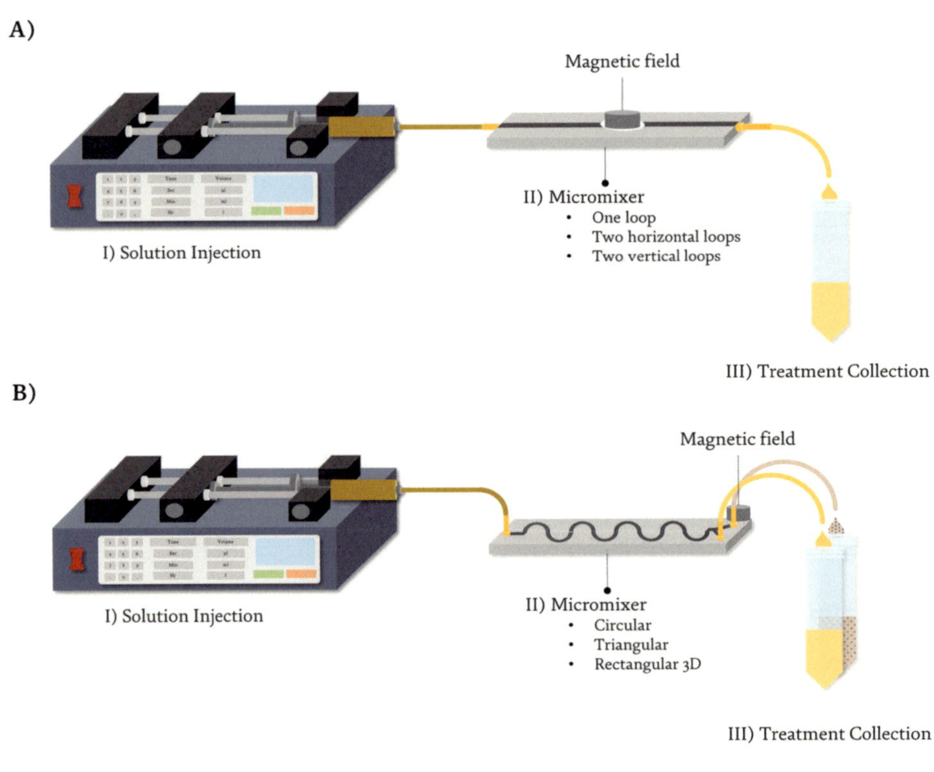

Figure 2.
Experimental scheme performed for on the micromixers by topologies with loops (a) and without (B).

349.23 mT, was externally inserted into the loops of some of the devices to retain the bionanocompounds while the reaction occurred. For devices that lack loops for permanent magnets, separation of the bionanocompounds by placing the magnets at the outlets of the system (See **Figure 2**). All experiments were carried out in triplicate. After the treatment, each sample was collected and analyzed spectrophotometrically in a GENESYS 10S UV–Vis v4.004 2L5R078128 (Thermo SCIENTIFIC, USA) An absorbance peak was monitored at 545 nm and also the absorbance area of the entire visible spectrum in the range between 400 and 700 nm was calculated. All measurements were carried out in triplicate.

2.6 LCA requirements

2.6.1 Goal and scope

This life cycle analysis aimed to evaluate possible impacts associated with manufacture and operation of six different micromixers for wastewater treatment. This LCA was based on an attributional approach or descriptive "cradle to gate"

Stage	Inventory	Amount	Unit
	Single layer micromixer		
	Polymethyl methacrylate (PMMA)	0.011	m^3
	Ethanol	1	ml
	Energy	0.339	kWh
	Water consumption	53	ml
	Two layers micromixer		
	Polymethyl methacrylate (PMMA)	0.017	m^3
Manufacturing	Ethanol	2	ml
	Energy	0.577	kWh
	Water consumption	51	ml
	Multiple layers micromixer		
	Polymethyl methacrylate (PMMA)	0.034	m^3
	Ethanol	5	ml
	Energy	1.293	kWh
	Water consumption	51	ml
	Enzyme activity assay		
	Citric acid	0.060	g
	Disodium hydrogen phosphate	0.050	g
	ABTS	0.110	g
	Energy	0.057	kWh
Operation	Water consumption	15	ml
	Dye preparation		
	Eriochrome black	0.0001	g
	Water consumption	5	ml
	Operation		
	Energy	0.020	kWh

Table 1.
Inventory report of micromixers manufacturing and operation.

of the laboratory-scale processes. The functional unit of this study was defined as 5 mL of treated wastewater by each micromixer. System boundaries were set from the use of raw materials for the manufacturing microfluidic devices and synthesis of Lac-Magnetite nanoparticles until the absorbance analysis of treated wastewater.

2.6.2 Life cycle inventory (LCI)

Data from the synthesis of Lac-magnetite and the process of wastewater treatment of each micromixer were measured on site. These data collection involved the determination of the relevant flows, use of reagents, emissions, wastes, and energy consumptions for this LCA study. Data concerning distribution of electricity and production of reagents were obtained from the Ecoinvent 3.6 database. Inventory report of this LCA study was mostly based on own laboratory experiments. **Table 1** shows the inventory report of raw materials, water consumption and energy required for the manufacturing of each micromixer and the corresponding operation process for wastewater treatment.

2.6.3 Impact assessment

Life cycle impact assessment (LCIA) aims to calculate the potential environmental and human health impacts associated with the manufacturing and operation of six micromixers for wastewater treatment. This LCIA was carried out with the aid of Ecoinvent 3.6 database. Characterization factors reported by the International Reference Life Cycle Data System (ILCD) method for LCIA were applied as impact assessment tools. In addition, eight impact categories were considered in this study: human toxicity non-cancer effects, human toxicity cancer effects, ecotoxicity freshwater, climate change total, resource depletion of minerals and metals, resource depletion of dissipated water, freshwater and terrestrial acidification, and photochemical ozone formation.

Regarding the assumptions, data concerning environmental impacts included the production of reagents necessary to synthetize the magnetite nanoparticles, i.e., the production of iron chlorides (II) and (III). However, environmental impact data to produce Tetramethylammonium Hydroxide (TMAH) has not been reported yet and therefore was neglected from the LCA analysis.

3. Results and discussion

3.1 Life cycle impact assessment of the manufacturing stage

The impact assessment was divided into two stages, the one related to raw materials and the manufacturing of micromixers, and the one that involved the operation in wastewater treatment. At the manufacturing stage, micromixers were analyzed based on the resources required for their fabrication and the number of PMMA layers to assemble them. For example, circular and triangular micromixers only required one PMMA layer, while one loop, two horizontal loops, and rectangular-3D micromixers were formed by two PMMA layers. Finally, the two vertical loops micromixer was formed by four or even more PMMA layers. Based on these features, the micromixers' manufacturing was analyzed individually in terms of environmental impacts. Alternatively, for the operation stage, each micromixer was analyzed based on its specific retention rate of Lac-magnetite in each work cycle. Finally, LCA results of both stages were added up to determine the total impact of each micromixer.

Figure 3 shows the detailed factors that, in the manufacturing stage of micromixers, impacted human health and global warming. These results showed that PMMA contributed with 55%, energy consumption with 44%, and other raw materials with only 1% of the total impact of the human toxicity, within the cancer effects category (see **Figure 3A**). Therefore, effects on human toxicity may be most likely associated with the use of PMMA for the manufacture of microfluidic devices. The selection of suitable materials to design micromixers with low costs and high manufacturability, is an important factor. New alternatives have been proposed to minimize environmental and human health impacts of PMMA. For instance, Wan et al. [17] demonstrated that PMMA used to fabricate microfluidic devices can be recycled multiple times preserving a high optical quality and their properties for biological experiments. Also, their results highlighted the importance of choosing appropriate parameters for the recycling process such as temperature, time, and pressure. Therefore, an alternative to reduce the impacts associated with the manufacture of micromixers in our case is by recycling the PMMA.

Figure 3B shows the impact assessment in climate change category. Energy consumption contributed with 53%, which can be explained by the energy consumed during the laser cutting process for the manufacture of micromixers. Overall, the results reflected that multiple layer micromixers showed the highest values on human toxicity and climate change categories, with up to 2 to 4 times increase in the values compared to two layers and single layer micromixers, respectively. This trend was also observed in other impact categories considered in this study. Although energy spent in the laser cutting had the highest contribution in the impact assessment, this technology reduced the micromixer manufacturing time compared to other wet chemical etching processes [22]. Also, this technique facilitated maintaining consistent dimensions and the appropriate device functionality due to its resolution and flexibility in terms of the variety of materials that can be handled [23]. Therefore, laser cutting offers significant benefits over other manufacturing techniques to achieve an accurate design of micromixers at very low cost.

Experimental tests determined that retention of Lac-magnetite nanoparticles was 87% for the two vertical loops micromixer, 80% for the one loop and the two horizontal loops micromixers, 40% for the triangular micromixer, and 0% for the rectangular-3D and circular micromixers. This analysis was carried out by measuring the amounts of Lac-magnetite bionanocompounds exiting the micromixer after wastewater treatment process. These nanoparticles remained attached to the walls of the micromixer in each work cycle. Based on retention information, the equivalent amount of Lac-magnetite held in each work cycle was estimated to calculate

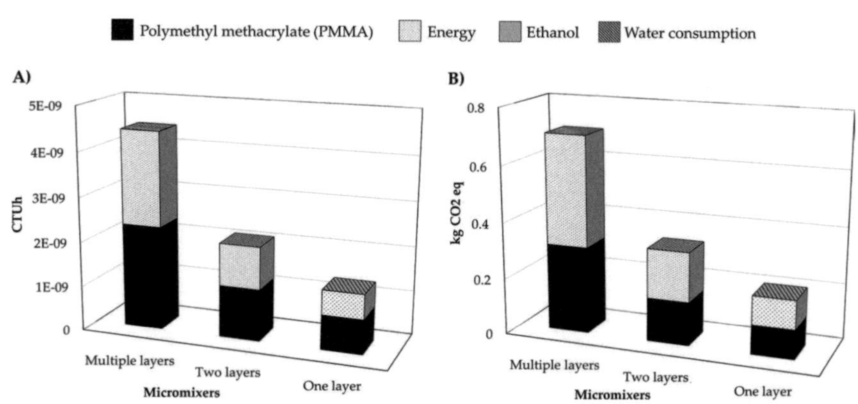

Figure 3.
Impact assessment in manufacturing stage for: A) cancer effects category of human toxicity, B) climate change.

the corresponding environmental impacts. Therefore, we considered the operation stage of each micromixer from its use in the first cycle until the completion of a total of ten work cycles. The initial input of the Lac-magnetite bionanocompound was 5 mg during all the operation process in the wastewater treatment. Then, this amount was different for each microsystem and work cycle. The total amounts of Lac-magnetite per work cycle for each micromixer are summarized in **Table 2**.

3.2 Life cycle impact assessment of the operation stage

Figure 4 shows the impact assessment results for the wastewater treatment operation stage for all micromixer devices. The total impact of each micromixer, in this stage, was determined by the summation of all impacts during ten work cycles. **Figure 4** compiles the results of all evaluated impact categories for the six devices under study. Overall, the LCIA results showed that in the operation stage, circular and rectangular-3D micromixers presented 30% more impact than the other micromixers. This finding can be explained by the high retention of these devices and the Lac-magnetite amount required for each work cycle. Two vertical loops and multiple layers micromixers presented the lowest impact in all impact categories, due to their high retentions per work cycle. Impact assessment was measured in four general categories: human health, ecosystem quality, climate change and resource depletion.

Figure 4A–C show impacts on human toxicity, non-cancer effects category, human toxicity-cancer effects category, and photochemical ozone formation category, respectively. Results indicate that human toxicity impacts are mainly related to ABTS use. This is a chemical compound used to track the reaction kinetics of specific enzymes such as laccases [24]. In this study, ABTS is used in the enzymatic activity assay of the obtained Lac-magnetite bionanocompounds. Assessment of toxicological effects of ABTS emitted into the environment were considered by estimating a specific characterization factor, i.e., comparative toxic units (CTUh). This factor provides an estimate of increase morbidity for the human population per unit mass of an emitted chemical (cases per kilogram) by assuming equal weighting between cancer and non-cancer situations [25]. However, some studies have

Micromixer		*Two vertical loops*	*One loop*	*Two horizontal loops*	*Triangular*	*Circular*	*Rectangular-3D*
Retention rate		*87%*		*80%*	*40%*		*0%*
Lac-magnetite amount (mg)	*Cycle 1*	5		5	5		5
	Cycle 2	0.65		1	3		5
	Cycle 3	0.0845		0.2	1.8		5
	Cycle 4	0.0109		0.04	1.08		5
	Cycle 5	0.0014		0.008	0.648		5
	Cycle 6	0.00018		0.0016	0.3888		5
	Cycle 7	2.4E-05		0.00032	0.2332		5
	Cycle 8	3.1E-06		6.4E-05	0.1399		5
	Cycle 9	4.08E-07		1.28E-05	0.0839		5
	Cycle 10	5.3E-08		2.5E-06	0.0503		5

Table 2.
Lac-magnetite amount per work cycle for the six micromixer devices.

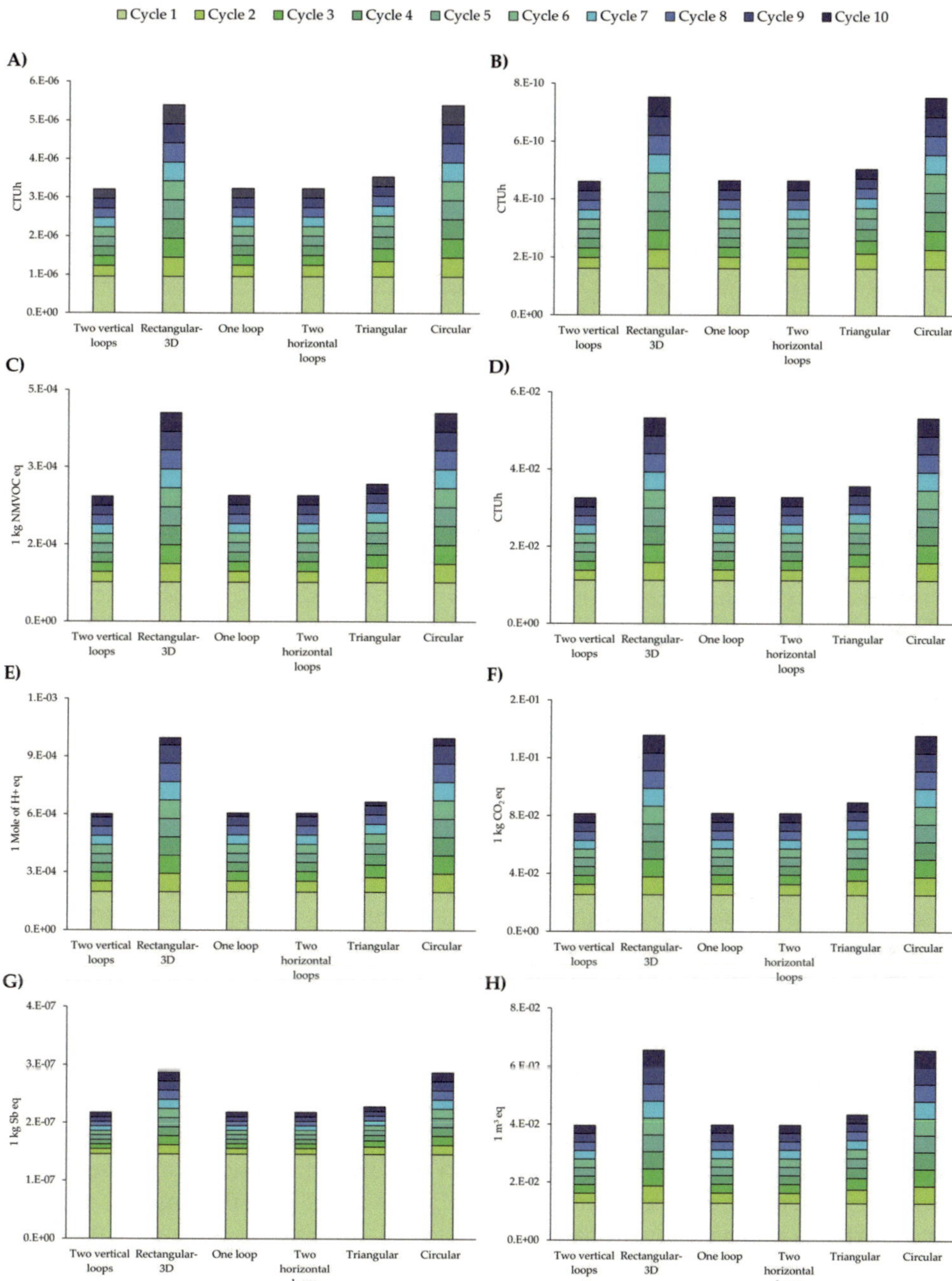

Figure 4.
Impact assessment results for the operation stage: A) human toxicity, non-cancer effects, B) human toxicity, cancer effects, C) photochemical ozone formation, D) Ecotoxicity freshwater, E) freshwater and terrestrial acidification, F) climate change, G) resource depletion, minerals and metals, H) resource depletion, dissipated water.

proposed the calculation of human health effect factors for cancer and non-cancer effects via ingestion and inhalation exposure, respectively. Additionally, toxic effects models have been considered to determine impacts on human health per kilogram substance emitted [26]. These calculations have been developed through steps, such as environmental fate, exposure, and effects of chemicals, which implies a cause–effect chain that links emissions to impacts.

Regarding the ecosystem quality, **Figure 4D** and **E** present impacts on the ecotoxicity freshwater category and the freshwater and terrestrial acidification category. According to Aurisano et al. [27], assessing ecotoxicological impacts on freshwater ecosystems after chemical exposure is an important component of various environmental and chemical management frameworks. These impact categories were considered here because we needed to determine impacts associated with compounds from our process that potentially come into contact with aquatic organisms and human beings [28]. Results showed that ABTS had the highest impact contribution on these categories due to its potential impact on aquatic ecosystems. Many authors have agreed that freshwater acidification is mainly caused by protons resulting from the mineralization of nitrogen and sulfur deposition, while carbon dioxide is the main cause of (coastal) marine acidification [29, 30]. These environmental impacts directly compromised the operation stage of micromixers in wastewater treatment.

Figure 4F shows impacts on the climate change category. Emissions of CO_2 and other greenhouse gases (GHGs), aerosols, and ozone precursors are thought to be responsible for detrimental climate impact [31]. In this study, energy use in operation processes of micromixers had the highest contribution to this impact category, which agrees well with previous studies [32]. This energy along with the energy used during the life span of a micromixer comprise the life-cycle energy and emissions footprint. According to Yousefi et al. [33], in addition to the energy consumption issue, greenhouse gas (GHG) emission issues and an understanding of emissions in a production process based on the kilogram of carbon equivalent (CO_2eq) are also critical in any production process. Several studies have reported some greenhouse gas removal technologies that will be needed to balance residual emissions and meet the emission targets [34]. Overall, most of these technologies proposed involve carbon dioxide removal or conversion of a higher global warming potential (GWP) gas to a lower GWP gas [35]. However, some removal technologies require significant amounts of energy for both installation and operation. Therefore, it is necessary to continue investigating in this field to assess potential environmental tradeoffs, including those related to energy use and climate change.

Finally, **Figure 4G** and **H** show the impact assessment results for the resource depletion of minerals and metals category and resource depletion of dissipated water category, respectively. Results in these categories are mainly associated with the energy consumption due to the use of non-renewables such as fossil fuels. According to Klinglmair et al. [36], resources could be evaluated according to their depletion (consumption related to geological or natural reserve), scarcity (economic availability) and their criticality (a resource that is scarce and crucial for society). Hence, depletion refers to the decrease of the physical amount of a resource that is available for future human use [37]. Minerals and metals depletion are considered within the abiotic depletion potential (ADP) method, which is recommended by the ILCD handbook and the Product Environmental Footprint (PEF) as the best available practice for assessing resource depletion on a midpoint level [37, 38]. Therefore, here we considered this impact category to determine the potential impacts associated with resource use when operating wastewater treatment processes enabled by the developed micromixers. However, both environmental and human health impacts related to extraction or use, such as toxic emissions, are kept as separate environmental impact categories, and resource depletion directly impacting ecosystem health was disregarded in importance.

3.3 Total impact assessment

Figure 5 shows the impact assessment results of manufacturing and operation stages for each micromixer. The manufacturing stage had the highest contribution

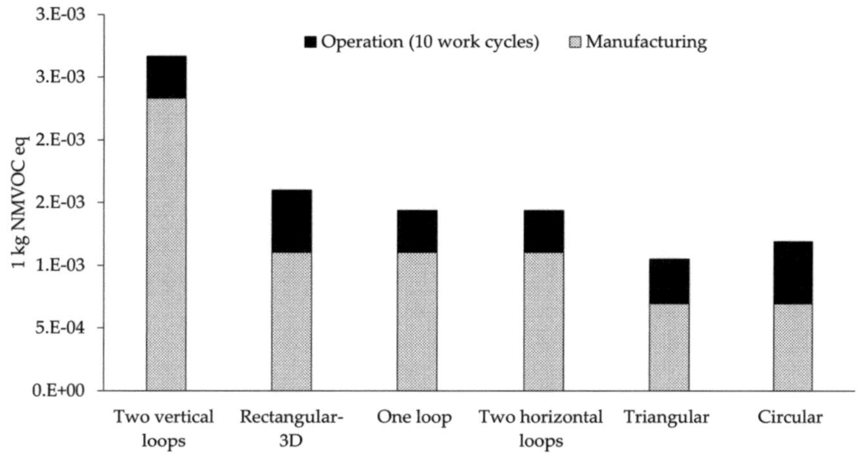

Figure 5.
Impact assessment of manufacturing and operation stages in the photochemical ozone formation category.

to the total impact of each micromixer in the photochemical ozone formation category, representing 87% in the two vertical loops micromixer, 69% in the rectangular-3D micromixer, 77% in the one loop and two horizontal loop micromixers, 66% in the triangular micromixer, and 58% the in circular micromixer. Similar results were also obtained in other impact categories due to the energy spent for laser cutting to manufacture the device in addition to the use of raw materials, such as PMMA. Regarding the operation stage, results showed that although ten work cycles for each micromixer were considered, this stage had the lowest contribution in all impact categories. This can be explained by the use of low impact raw materials in the enzyme activity assay, the preparation of artificial wastewater, and the micromixer operation.

Specifically, the two vertical loops micromixer presented, on average, 56% more impact in all categories than other micromixers, considering the manufacturing and operation stages. In contrast, the circular micromixer had the lowest impact in the manufacturing stage due to a significant reduction in the use of PMMA. Also, this micromixer had the highest impact during the operation stage due to its low retention of Lac-Magnetite, which leads to an increased requirement of the bionanocompound per cycle. However, total impact of circular micromixer is one of the lowest compared to other designs. This result showed that to calculate the impact assessment, it is necessary to consider all stages of a micromixer from its manufacture to its final operation.

4. Conclusions

Results from this study showed that six prototypes of micromixers for wastewater treatment can be analyzed in terms of impacts to human health and environment using the LCA methodology. This tool confirmed to be useful for this early research stage as it allows to identify potential impacts during the different phases required to implement these technologies.

According to the four general impacts categories considered in this study, we successfully identified the main flows that contributed to each one. The ABTS chemical for enzyme activity assays significantly contributed to human health and ecosystem quality categories. Assessment of potential toxicological effects of this compound on human health were determined in several impact categories including human toxicity, cancer and non-cancer effects, and photochemical ozone production. Also, in terms of ecosystem quality, impacts on the ecotoxicity freshwater and the freshwater and terrestrial acidification categories were considered. Toxic effects of the ABTS were the highest compared to other raw materials during the operation stage mainly due to its release to aquatic ecosystems where it might eventually reach organisms and human beings. Moreover, energy use contributed to climate change and resource depletion categories. Emissions of CO_2 and other greenhouse gases were considered in the climate change category. Regarding the resource depletion category, results showed that the use of non-renewables such as fossil fuels to produce electricity was the major contributor to this category.

Multiple layers micromixers showed the highest impact while the one-layer ones the lowest. These results were associated directly with the manufacturing stage, where PMMA and energy used had the highest contribution to impacts on environmental and human health categories, respectively. Therefore, the manufacturing stage had the highest contribution to the total impact of each micromixer in all impact categories. Also, the operation stage depended directly on the retention of the active bionanonanocompounds within each micromixer, in addition to others raw materials necessary for wastewater treatment. Finally, impact assessment results of the manufacturing and operation stages determined the total impact of a micromixer during its work cycle.

This study represents a first step for the impact assessment on the environment and human health of the wastewater bioremediation treatment enables by low and high efficiency micromixers. Moreover, this work sets a starting point to further explore the potential of micromixers and the possible environmental concerns arising from their implementation in large-scale operations.

Acknowledgements

O.P. Fuentes is funded by the Ministerio de Ciencia, Tecnología e Innovación (Colombia) with the grant 766 of 2016. Authors also want to thank Universidad de los Andes for the economic support of this work.

Author details

Olga P. Fuentes[1], Mabel J. Noguera[1], Paula A. Peñaranda[1], Sergio L. Flores[1], Juan C. Cruz[2] and Johann F. Osma[1*]

1 Department of Electrical and Electronic Engineering, Universidad de los Andes, Bogotá D.C., Colombia

2 Department of Biomedical Engineering, Universidad de los Andes, Bogotá D.C., Colombia

*Address all correspondence to: jf.osma43@uniandes.edu.co

References

[1] Tabeling P, Chen S. Introduction to Microfluidics. OUP Oxford; 2010. 301 p. ISBN 9780199588169

[2] Nguyen N, Wereley S, Mousavi S. Fundamentals and Applications of Microfluidics. Third Edition: Artech House; 2018. 576 p. ISBN: 9781630813642

[3] Ahn CH., Choi JW. Microfluidics and Their Applications to Lab-on-a-Chip. In: Bhushan B. Springer Handbook of Nanotechnology. Springer Handbooks. Springer, Berlin, Heidelberg; 2004. p. 253-278. DOI: 10.1007/3-540-29838-X_9

[4] Ma J, Yi C, Li, CW. Facile synthesis and functionalization of color-tunable Ln3+–doped KGdF4 nanoparticles on a microfluidic platform. Materials Science and Engineering C. 2020; 108: 1-8. DOI: 10.1016/j.msec.2019.110381

[5] Minteer SD, Moore CM. Overview of Advances in Microfluidics and Microfabrication. In: Minteer SD. Methods in Molecular Biology™: Humana Press; 2006. p. 1-2. DOI: 10.1385/1-59259-997-4:1

[6] Farshchi F, Hasanzadeh M. Biomedicine & Pharmacotherapy Microfluidic biosensing of circulating tumor cells (CTCs): Recent progress and challenges in efficient diagnosis of cancer. Biomed Pharmacother. 2021;134:111153. DOI: 10.1016/j.biopha.2020.111153

[7] Kung C, Gao H, Lee C, Wang Y, Dong W. Microfluidic synthesis control technology and its application in drug delivery, bioimaging, biosensing, environmental analysis and cell analysis. Chem Eng J. 2020;399:125748. DOI: 10.1016/j.cej.2020.125748

[8] Weibel DB, Whitesides GM. Applications of microfluidics in chemical biology. Current Opinion in Chemical Biology. 2006;584-591. DOI: 10.1016/j.cbpa.2006.10.016

[9] Cleary J, Maher D, Diamond D. Development and Deployment of a Microfluidic. In: Mukhopadhyay S, Mason A, editors. Smart Sensors for Real-Time Water Quality Monitoring. Springer, Berlin, Heidelberg; 2013. p. 125-148. DOI: 10.1007/978-3-642-37006-9_6

[10] Khatri A, Hussain M, Mohsin M, White M. A review on developments in dyeing cotton fabrics with reactive dyes for reducing effluent pollution. J Clean Prod. 2015;87:50-57. DOI: 10.1016/j.jclepro.2014.09.017

[11] Yew M, Ren Y, Koh KS, Sun C, Snape C. A Review of State-of-the-Art Microfluidic Technologies for Environmental Applications: Detection and Remediation. Global Challenges. 2019;3:1800060. DOI: 10.1002/gch2.201800060

[12] Maguire I, O'Kennedy R, Ducrée J, Regan F. A review of centrifugal microfluidics in environmental monitoring. Analytical Methods. 2018;10:1497-1515. DOI: 10.1039/C8AY00361K

[13] Moreira NS, Chagas CLS, Oliveira KA, Duarte-junior GF, de Souza FR, Santhiago M, Garcia CD, Kubota LT, Coltro WKT. Fabrication of microwell plates and microfluidic devices in polyester films using a cutting printer. Analytica Chimica Acta. 2020;1119:1-10. DOI: 10.1016/j.aca.2020.04.047

[14] Faustino V, Catarino SO, Lima R, Minas G. Biomedical microfluidic devices by using low-cost fabrication techniques: A review. Journal of Biomechanics. 2016;49: 2280-2292. DOI: 10.1016/j.jbiomech.2015.11.031

[15] Jinfang N, Yuanzhi L, Yun Z, Shangwang L, Dunnan L, Songbai Z. One-step patterning of hollow microstructures in paper by laser cutting to create

microfluidic analytical devices. Analyst. 2013;138:671-676. DOI: 10.1039/C2AN36219H

[16] Gale BK, Jafek AR, Lambert CJ, Goenner BL, Moghimifam H, Nze UC, Kamarapu SK. Review of Current Methods in Microfluidic Device Fabrication and Future Commercialization Prospects. Inventions. 2018;3:60. DOI: 10.3390/inventions3030060

[17] Wan AMD, Devadas D, Young EWK. Recycled polymethylmethacrylate (PMMA) microfluidic devices. Sensors Actuators B Chem. 2017;253:738-744. DOI: 10.1016/j.snb.2017.07.011

[18] Pradel M, Garcia J, Vaija MS. A framework for good practices to assess abiotic mineral resource depletion in Life Cycle Assessment. J Clean Prod. 2021;279:123296. DOI: 10.1016/j.jclepro.2020.123296

[19] Corominas L, Byrne DM, Guest JS, Hospido A, Roux P, Shaw A, et al. The application of life cycle assessment (LCA) to wastewater treatment: A best practice guide and critical review. Water Research. 2020;184:116058. DOI: 10.1016/j.watres.2020.116058

[20] Lopes TAS, Queiroz LM, Torres EA, Kiperstok A. Low complexity wastewater treatment process in developing countries: A LCA approach to evaluate environmental gains. Science of the Total Environment. 2020;720:137593. DOI: 10.1016/j.scitotenv.2020.137593

[21] Ramírez-Cavazos LI, Junghanns C, Ornelas-Soto N, Cárdenas-Chávez DL, Hernández-Luna C, Demarche P, et al. Purification and characterization of two thermostable laccases from Pycnoporus sanguineus and potential role in degradation of endocrine disrupting chemicals. J Mol Catal B Enzym. 2014;108:32-42. DOI: 10.1016/j.molcatb.2014.06.006

[22] Bermudez JF, Saldarriaga JF, Osma JF. Portable and Low-Cost Respirometric Microsystem for the Static and Dynamic Respirometry Monitoring of Compost. Sensors. 2019;*19*:4132. DOI: 10.3390/s19194132

[23] Gale BK, Jafek AR, Lambert CJ, Goenner BL, Moghimifam H, Nze UC, Kamarapu SK. A Review of Current Methods in Microfluidic Device Fabrication and Future Commercialization Prospects. Inventions. 2018;3:60. DOI: 10.3390/inventions3030060

[24] Leng Y, Bao J, Xiao H, Song D, Du J, Mohapatra S, et al. Transformation mechanisms of tetracycline by horseradish peroxidase with/without redox mediator ABTS for variable water chemistry. Chemosphere. 2020;258:127306. DOI: 10.1016/j.chemosphere.2020.127306

[25] Rosenbaum RK, Bachmann TM, Gold LS, Huijbregts MAJ, Jolliet O, Juraske R, et al. USEtox - The UNEP-SETAC toxicity model: Recommended characterisation factors for human toxicity and freshwater ecotoxicity in life cycle impact assessment. Int J Life Cycle Assess. 2008;13(7):532-546. DOI: 10.1007/s11367-008-0038-4

[26] Rosenbaum RK, Huijbregts MAJ, Henderson AD, Margni M, McKone TE, Van De Meent D, et al. USEtox human exposure and toxicity factors for comparative assessment of toxic emissions in life cycle analysis: Sensitivity to key chemical properties. Int J Life Cycle Assess. 2011;16(8):710-727. DOI: 10.1007/s11367-011-0316-4

[27] Aurisano N, Albizzati F, Hauschild M, Fantke P. Extrapolation Factors for Characterizing Freshwater Ecotoxicity Effects. Environ Toxicol Chem. 2019;38(11):2568-2582. DOI: 10.1002/etc.4564

[28] Arvidsson R, Foss S, Baun A. Influence of natural organic matter on

the aquatic ecotoxicity of engineered nanoparticles: Recommendations for environmental risk assessment. NanoImpact. 2020;20:100263. DOI: 10.1016/j.impact.2020.100263

[29] Van Zelm R, Roy PO, Hauschild MZ, Huijbregts MAJ. Acidification. In: Hauschild M., Huijbregts M, editors. Life Cycle Impact Assessment. LCA Compendium – The Complete World of Life Cycle Assessment. Springer, Dordrecht. 2015. p. 163-176. DOI: 10.1007/978-94-017-9744-3

[30] Zak D, Hupfer M, Cabezas A, Jurasinski G, Audet J, Kleeberg A, et al. Sulphate in freshwater ecosystems: A review of sources, biogeochemical cycles, ecotoxicological effects and bioremediation. Earth-Science Reviews. 2021;212: 103446. DOI: 10.1016/j.earscirev.2020.103446

[31] Levasseur A, Cavalett O, Fuglestvedt JS, Gasser T, Johansson DJA, Jørgensen S V, et al. Enhancing life cycle impact assessment from climate science: Review of recent findings and recommendations for application to LCA. Ecol Indic. 2016;71:163-174. DOI: 10.1016/j.ecolind.2016.06.049

[32] Li DHW, Yang L, Lam JC. Impact of climate change on energy use in the built environment in different climate zones - A review. Energy. 2012;42(1):103-112. DOI: 10.1016/j.energy.2012.03.044

[33] Yousefi M, Damghani AM, Khoramivafa M. Comparison greenhouse gas (GHG) emissions and global warming potential (GWP) effect of energy use in different wheat agroecosystems in Iran. Environ Sci Pollut Res. 2016;23:7390-7397. DOI: 10.1007/s11356-015-5964-7

[34] P. Williamson. Emissions reduction: scrutinize CO2 removal methods. Nature. 2016;530:153-155. DOI: 10.1038/530153a

[35] Goglio P, Williams AG, Balta-ozkan N, Harris NRP, Williamson P, Huisingh D, et al. Advances and challenges of life cycle assessment (LCA) of greenhouse gas removal technologies to fight climate changes. J Clean Prod. 2020;244:118896. DOI: 10.1016/j.jclepro.2019.118896

[36] Klinglmair M, Sala S, Brandão M. Assessing resource depletion in LCA: A review of methods and methodological issues. Int J Life Cycle Assess. 2014; 19(3):580-592. DOI: 10.1007/s11367-013-0650-9

[37] Schneider L, Berger M, Finkbeiner M. Abiotic resource depletion in LCA—background and update of the anthropogenic stock extended abiotic depletion potential (AADP) model. Int J Life Cycle Assess. 2015;20(5):709-721. DOI: 10.1007/s11367-015-0864-0

[38] van Oers L, Guinée JB, Heijungs R. Abiotic resource depletion potentials (ADPs) for elements revisited—updating ultimate reserve estimates and introducing time series for production data. Int J Life Cycle Assess. 2020;25(2):294-308. DOI: 10.1007/s11367-019-01683-x

Chapter 7

Solar Thermal Conversion of Plasmonic Nanofluids: Fundamentals and Applications

Meijie Chen, Xingyu Chen and Dongling Wu

Abstract

Plasmonic nanofluids show great interests for light-matter applications due to the tunable optical properties. By tuning the nanoparticle (NP) parameters (material, shape, and size) or base fluid, plasmonic nanofluids can either absorb or transmit the specific solar spectrum and thus making nanofluids ideal candidates for various solar applications, such as: full spectrum absorption in direct solar absorption collectors, selective absorption or transmittance in solar photovoltaic/thermal (PV/T) systems, and local heating in the solar evaporation or nanobubble generation. In this chapter, we first summarized the preparation methods of plasmonic nanofluids, including the NP preparation based on the top-down and bottom-up, and the nanofluid preparation based on one-step and two-step. And then solar absorption performance of plasmonic nanofluids based on the theoretical and experimental design were discussed to broaden the absorption spectrum of plasmonic nanofluids. At last, solar thermal applications and challenges, including the applications of direct solar absorption collectors, solar PT/V systems, solar distillation, were introduced to promote the development of plasmon nanofluids.

Keywords: solar thermal, plasmonic, nanofluid, absorption, nanoparticle

1. Introduction

Nowadays, with the development of society and the improvement of people's living standards, environment issues, such as: greenhouse effect, acid rain, and haze, has become more serious. Developing green energy technology has attracted more researchers' attention, especially for solar energy, which is universal, harmless, huge, and sustainable. Solar utilizations can be divided into two main categories: solar-electric and solar-thermal. And both of them are needed to enhance the solar absorption performance of working media at its first step of solar conversion applications.

Solar thermal conversion is one of the most simple and direct ways of solar utilizations by heating the working mediums directly for follow-up usages, which can be widely used in the solar thermal collectors [1], solar distillation [2, 3] and so on [4]. Therefore, it's critical to improve the solar absorption performance of working mediums for the solar thermal conversion applications. For example, based on the surface absorber, various nanostructure coatings (e.g., grating, porous structure, and so on [5, 6]) were designed to achieve the selective absorption ability, which serves as solar selective absorbers by heating the surface and transferring the heat to the working fluid for the follow-up applications. The heat loss from the absorbed surface due to the high temperature and heat transferred resistance between the absorbed surface and working should be considered during the design processes, which also limits its large-scale practical application at the high or middle temperature solar thermal conversion applications.

Instead of absorbing solar energy by a surface, work fluid can be used to absorb solar energy, which serves as both the solar absorber and heat transfer medium and can avoid the local high temperature area and reduce the heat transfer resistance. However, the common working fluids such as: water, oil, and alcohol usually have the limited solar absorption ability [7]. It was found that adding nanoparticles (NPs) to these working fluid (i.e., nanofluid) can greatly improve the solar collector efficiency [8, 9]. Nanofluid is a suspension of NPs (1–100 nm) in a conventional base fluid, which was first used by Choi in 1995 [10]. Nanofluids show unique characteristics in many aspects, including the heat transfer [11, 12] and the solar absorption ability due to the interaction between the light and NPs at nanoscale [9, 13]. For example, carbon nanotube, graphite and the other black carbon NPs were added into the base fluid to achieve the great solar absorption performance [14].

Plasmonic nanofluids show great interests to improve the absorption ability by dispersing plasmonic NPs in the base fluid stability. Due to the surface plasmon resonance (SPR) around the NP surface [15], the incident electric coupled with the free electron oscillation around the NP surface at the resonance frequency can strongly enhance the absorption performance of NPs [16] in **Figure 1**. The optical absorption performance of nanofluids can be enhanced by tuning the NP shape, size, or base fluid. Using plasmonic nanofluids as the absorber and heat transfer medium in the solar thermal applications shows great potential due to the excellent optical and thermal characteristics. To choose a proper nanofluids for specific solar thermal applications (such as: solar collectors, solar PV/T systems), many researchers investigated the optical and thermal properties of various nanofluids. For example, for the direct absorption solar collectors (DASCs), nanofluids as the absorber need to absorb the solar radiation in the full solar spectrum (0.3–2.5 um). While the nanofluid only serves as a beam splitter (i.e., selective absorber) in solar PV/T systems, which absorbs the useless spectrum for the PV cell and avoids heating the PV cells to improve the overall PV/T efficiency [4]. Hence, the optical absorption performance of plasmonic nanofluids should be considered in different solar thermal applications.

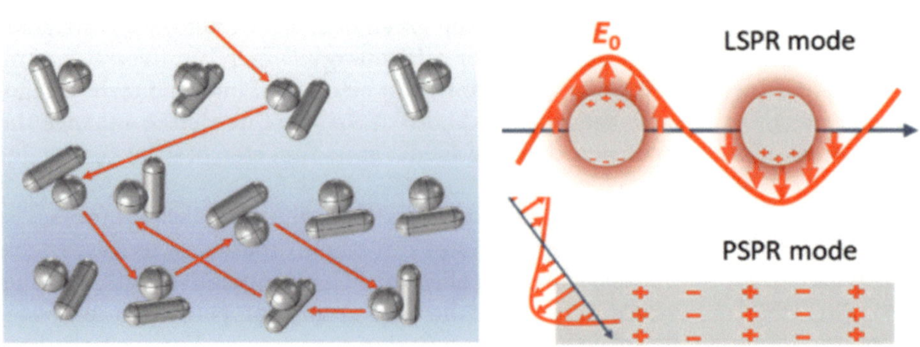

Figure 1.
Light propagation in the nanofluid [17] and the surface plasmon resonance (SPR) around the NP surface, dividing into localized and propagating surface plasmon resonance (LSPR and PSPR).

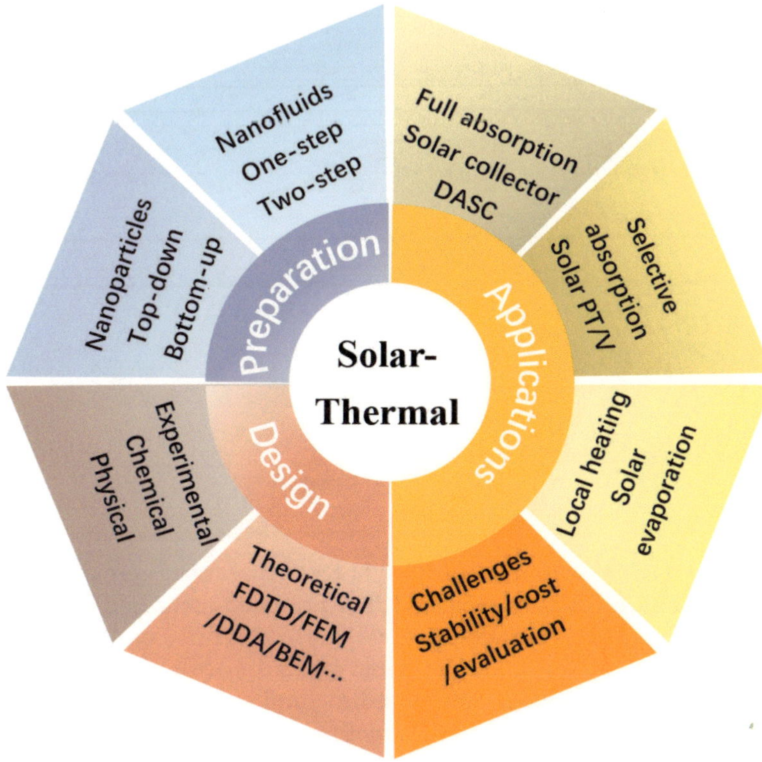

Figure 2.
The main parts of this chapter for solar thermal conversion of plasmonic nanofluids.

In this chapter, we focus on the solar thermal conversion of plasmonic nanofluids in **Figure 2**, which consists of the following three parts: 1) plasmonic nanofluid preparation including NPs and nanofluids; 2) solar absorption of plasmonic nanofluids based on the theoretical and experimental design; 3) solar thermal applications and challenges, including direct solar absorption collectors, solar PT/V systems, solar evaporation, other applications and challenges. To increase the understanding of previous studies, related analyses and calculation techniques are illustrated. This chapter is expected to provide researchers with deep insight into the solar thermal conversion of plasmonic nanofluids and facilitate future studies in this field.

2. Plasmonic nanofluid preparation

As discussed above, the NP parameters and dispersed base fluid have great effect on the optical absorption and solar thermal conversion performance of plasmonic nanofluids. We will first discuss the preparation methods of plasmonic nanofluids, and then the preparation methods of plasmonic NPs (**Figure 3**) are summarized due to the great interaction of NP parameters with the light. In this section, some common methods to prepare plasmonic NPs or nanofluids are listed and their advantages or limitation would also be discussed.

2.1 Plasmonic nanofluid preparation

The preparation method of nanofluids can be classified into two main categories in **Figure 3a**: one-step method and two-step method [18].

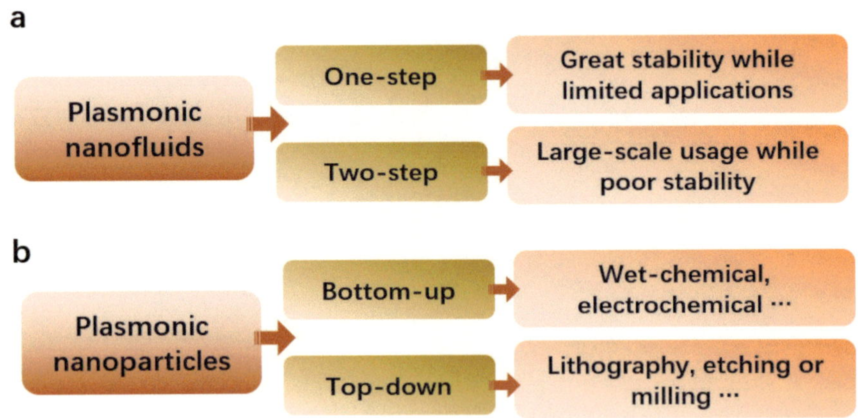

Figure 3.
Preparation methods of (a) plasmonic nanofluids and (b) nanoparticles.

The one-step method is to disperse NPs in the NP synthesis process by using the physical method or wet-chemical method. The prepared nanofluid is relatively stable by avoiding the NP separation process (e.g., centrifuge or drying) and redisperse process (e.g., stirring or ultrasonic oscillation). Hence, the one-step method can reduce NP agglomeration or sedimentation in the nanofluids. For example, an ultrasound-assisted one-step method was used to prepare spherical and plate-shaped Au NPs with the NP size of 10 $\sim$ 300 nm [19]. But the dispersed base fluid has limitations, which usually consists of the residual chemical reagent in the NP synthesize process and the nanofluid could not be applied in a large-scale range.

The two-step method separates the NP synthesis process from the dispersion process, which is widely used in the large-scale applications. In this method, various NPs, such as: nanospheres, nanorods, nanotubes, and so on, are in the state of dry powders and then these NPs can be dispersed into different base fluids by stirring or ultrasonic oscillation for different applications. Due to the redispersion process of NPs, the stability is worse than that of the one-step method, but the two-step method has a simple preparation process, a high NP controllability, and wide application ranges. It can be seen from the two-step method that the NP parameters, such as the size, morphology, and dielectric environment of the base fluid determine their unique SPRs. Therefore, for solar thermal conversion applications of plasmonic nanofluids, preparing plasmonic NPs with the controllable morphology and size is the prerequisite and basis for their applications. And the fabrication of these plasmonic NPs will be discussed in the next section.

2.2 Plasmonic nanoparticle fabrication

The fraction processes of plasmonic NPs can be divided into two main categories in **Figure 3b**: the top-down based on the lithography, etching or milling, and the bottom-up, including the seed-mediated growth, chemical reduction, electrochemical method, and so on. Given that the NP shape can significantly affect the way it interacts with light and its SPRs, researchers have made great efforts to develop preparation methods for plasmonic NPs with the reproducible control of the size and shape. Nowadays, it's possible to fabricate high-quality plasmonic NPs (e.g., Au or Ag) with the target SPR wavelengths or near-field enhancement by enabling a systematic study of SPR dependencies on the size, shape, and structure of plasmonic NPs.

The wet-chemical method is one of the most common bottom-up methods to tune the size or shape of metal NPs. For Au NPs, chloroauric acid ($HAuCl_4$) in aqueous solution is usually used as the precursor and the reducing agent can be polyvinyl pyrrolidone (PVP), Sodium borohydride ($NaBH_4$), ascorbic acid (AA), sodium citrate (SC), and so on. The reduction reaction is also determined by the temperature. And various Au NP shapes can be achieved by introducing the additive, e.g., etyltrimethylammonium bromide (CTAB) for nanorods [20], sodium citrate or trisodium citrate for spheres [21], silver iron (Ag^+) for thorns [22]. Au NPs with the size ranged from 5 nm to 150 nm can be obtained by changing the concentration of the precursor and the reducing agent [23], PH value [24], temperature [25] and so on [26]. Seed-medicated method is another effective method to control the size and shape of Au NPs, which divides the reaction into nucleation and growth stages separately, so that the size and morphology of the particles can be controlled in a larger range [27]. By controlling the growth rate of different crystal planes, the final synthesized NPs can deviate from the initial seed crystal structure by the seed growth method. For example, cube Au NPs were prepared by the amount of ascorbic acid (AA) in the growth solution and the size can be controlled by the amount of seed solution or the growth number [28, 29]. Ag NPs also can be prepared by the similar methods using $AgNO_3$ or other silver salts. To scalable and green prepare metal NPs, a rotating electrodeposition and separation (REDS) technique developed, which entails electrochemically depositing NPs onto a continuously rotating metal foil and subsequently harvesting them through mechanical delamination. A wide array of elemental nanoparticles (e.g., Ag, Au, Ni, Cu), alloys nanoparticles (e.g., FeCoNi and FeCoNiW), and metal oxide nanomaterials (e.g., Co_3O_4) were synthesized by REDS [30].

Besides the bottom-up method, top-down method is another to prepare NPs, which mainly including some physical method, such as: electron beam lithography (EBL), milling, annealing, laser-melting, and so on. The fabrication process of EBL is similar to classical photolithography, an electron beam is used to mark the pattern in the resist instead of light. [31]. Despite the low fabrication throughput and the fact that very small structures may be at the physical limit in terms of electronic function, the technique can find applications in preparation of reproducible large-scale arrays of plasmonic NP with arbitrary two-dimensional shapes. Among the disadvantages are the technological requirements such as high vacuum and a scanning electron microscope system, much longer time to write a pattern than photolithography. Using conventional lithography techniques many shapes of nanoparticles on surfaces can be achieved. However, large scale fabrication using reproducible patterns, inverse replication or transfer of NPs between substrates, and three-dimensional nanostructures including deep etching has been increasingly demanded in plasmonic nanostructures [32, 33].

After obtaining different plasmonic NPs, it should be dispersed into the working fluid stability to form various nanofluids. The fabrication of NPs based on the top-down method is expensive and consumed on the materials, which could not meet the scale requirement although the high-quality NPs can be achieved. Wet-chemical method can be an efficient method to achieve plasmonic NPs in the solar thermal conversion applications.

2.3 Nanofluid stability

Nanofluid is defined as dispersing NPs stability into the base fluid, which is not simply mixing solid NP phase and liquid phase. It's a complex colloid by dispersing specific functional NPs in the base fluid (e.g., water, oil and so on). The main challenge for nanofluid applications is how to produce well-dispersion nanofluids.

Owing to the interaction among different NPs at the nanoscale and gravity at Earth, NPs are usually agglomerated due to Van der Waals force and then trend to be sediment at the bottom [34]. As a result, the agglomeration and sedimentation of NPs in the base fluid would affect the optical absorption and heat transfer performance, weakening the system efficiency. In addition, recent studies showed that the agglomeration or sedimentation can be worse under harsh operating conditions, such as: high temperature and pressure [35, 36]. Many methods were used to evaluate the stability of nanofluids, the simplest and direct method is the sedimentation method [18]. Interface electromotive force analysis is another common method to observe the stability of nanofluids, but this method is limited by the viscosity and concentration of the fluid [37]. Wang et al. [38] used an ultraviolet–visible spectrophotometer to study the stability of nanofluids. The NP concentration can be obtained by measuring the change in the light absorption rate of the system with the sedimentation time because the NP concentration is a linear relationship with the absorbance of nanofluid at the low concentration.

The stability of plasmonic nanofluids is also one of the major issues limiting the applications of nanofluids. Many researchers have made much efforts to improve the stability of the plasmonic nanofluids from the aspect of long-time and high-temperature dispersion [39]. For example, Au@SiO_2 and Ag@SiO_2 core-shell NPs were synthesized using a low-temperature two-step solution process. Results showed that the synthesized metal@SiO_2 nanofluids exhibited excellent dispersion stability of 93.7% for Au@SiO_2 and 100% for Ag@SiO_2 in 6 months without using any surfactants, and they also showed a good thermal stability after thermal exposure at 150° C for an hour [40]. An ultrastable nanofluids with the broadband photothermal absorption was achieved using citrate and polyethylene glycol-coated Au NPs, circumventing the need for free surfactants. Electrostatic stabilization provided superior colloidal stability and more consistent optical properties; chemical and colloidal stability was verified for 16 months, the longest demonstration of stable nanofluids under ambient storage in the solar literature [41]. Besides the base fluid water used above, the base fluid oil was also studied to improve the stability. A facile and effective strategy, including controlled high-temperature synthesis of nanoparticles, surface modification of particles, and post-modification particle size partition, was designed to prepare stably dispersed silicone-oil-based nanofluids that enable high-temperature operation [42]. A low cost, and scalable method was reported to synthesize solar selective nanofluids from 'used engine oil' with the excellent long-term stability and photothermal conversion efficiency. Results showed that their stability and functional characteristics can retain even after extended periods (72 hours) of high temperature (300°C) heating, ultra violet light exposure and thermal cyclic loading [43].

3. Solar absorption of plasmonic nanofluids

The excellent optical absorption performance of plasmonic NPs make it to be a great candidate in the solar thermal conversion applications, which is critical for the solar thermal conversion applications. And the optical properties of nanofluids can be controlled by the NP size, shape, concentration and base fluid. In this section, we will discuss the optical properties of NPs or nanofluids from two aspects: theoretical design and experimental design.

3.1 Theoretical design

To achieve the optical properties of nanofluids, including transmittance, reflectance, and absorptance, the optical performance of single NP is usually determined firstly. And the dielectric function of materials are required for the optical simulation, which is taken from the experimental data of bulk materials (e.g., Johnson and Christy [44],) or a model approximating experimental results (e.g., Drude method). The Drude model is the simplest of all, but disregards radiation damping. Even today, mainly because of the simplicity, the Drude model is still used to describe the dielectric functions in many calculations. In some problems, the classical models of dielectric functions are unsatisfactory but, at the same time, full quantum theories involve a very complex treatment including non-local effects [45], polarizabilities including non-linear terms [46], electron densities calculation using mean-field theories [47] and temperature dependent effects [48]. The need for quantum treatment of the optical properties of small particles has been evidenced in recent experimental studies [49]. In large particles the resonances are influenced by retardation effects and are strongly dependent on the size of particles, but the dielectric function can be assumed as that of bulk. Based on the dielectric function, Au, Ag and Al are the three most used materials in plasmonics. Their SPR wavelengths are at visible or UV spectral bands and, therefore, of great potential in solar thermal applications.

Mie theory: Mie theory is a simple and theoretical method to calculate the optical properties of sphere NPs in a homogenous medium, which uses a series of coefficients a_n and b_n for the scattered fields and c_n and d_n for the internal fields to determine the scattering fields. The scattering and extinction cross sections can be calculated as: [50].

$$C_{scat} = \frac{2\pi}{k^2}\sum_{n=1}^{\infty}(2n+1)\left(|a_n|^2 + |b_n|^2\right) \tag{1}$$

$$C_{ext} = \frac{2\pi}{k^2}\sum_{n=1}^{\infty}(2n+1)\,\mathrm{Re}\,(a_n + b_n) \tag{2}$$

The absorption cross section can be obtained as: $C_{abs} = C_{ext} - C_{scat}$. Despite to the less computation load, it is possible to obtain cross-sections for many wavelengths in a few seconds, using a common PC. However, a large number of terms is required for accurate cross-section calculations of spheres with very large size parameter [51]. The Mie theory has been extended to permit calculations for ellipsoidal shape, multilayer or several spheres [52].

DDA: To calculate the light scattering of an arbitrary shape NP, discrete dipole approximation (DDA) was first presented by Purcell and Pennypacker [53] by using a grid of dipoles. To occupy by the scattering target, DDA method discretizes the volume by an array of N dipoles using Clausius–Mossotti polarizability α_j for each dipole, which interacts with the incident field and the neighbors. The polarization of dipole j located at r_j can be determined by $P_j = \alpha_j E_j$, and the field can be calculated as:

$$E_j = E_j^{inc} - \sum_{k \neq j} A_{jk} P_k \tag{3}$$

where $E_j^{inc} = E_0 e^{ikr - i\omega t}$. A_{jk} is the matric of dipole interaction and retardation effect.

To achieve accurate and reproduce the calculation results, two validity conditions should be verified in DDA: (a) the dipole lattice spacing d should be small enough, i.e., $|m|kd \leq 1$, where m is the complex refractive index of the scattering target. (b) d must be small enough to refabricate accurately the NP shape. For small plasmonic NPs, or small inter-particle separations, d must be smaller than 1 nm.

BEM: Boundary element method (BEM) is another method to calculate the optical properties of plasmonic nanostructures, which was introduced by García de Abajo and Howie [54] using the following equations:

$$\phi(r) = \phi_j^{ext}(r) + \int_{S_j} G_j(|r-s|)\sigma_j(s)ds \tag{4}$$

$$A(r) = A_j^{ext}(r) + \int_{S_j} G_j(|r-s|)h_j(s)ds \tag{5}$$

where $\phi(r)$ is the electric potential, $A(r)$ is the vector potential, $\sigma_j(r')$ is the surface charge density and $h_j(r')$ is the surface current density and G_j is the Green's function of Helmholtz equation inside each homogeneous medium of dielectric function. S_j is the boundary of the medium j. s is the point of the boundary between medias. r is the point inside the medium. $\phi^{ext}(r)$ and $A^{ext}(r)$ are the potentials at the interface caused by external sources and the full space is filled by a homogeneous medium j. Therefore, a much smaller number of elements is required to evaluate the fields than volume integral based methods, but involving a complex parameterization of the boundary elements.

FDTD: Finite-difference time-domain (FDTD) is one of the most popular optical calculation methods in plasmonic nanostructures, which is first developed by Yee in 1966 [55]. The basis of the model is from Maxwell equations in electrodynamics. The second-order precision central difference is used to approximate the discretization of the differential form of Maxwell equations, thereby a set of time-domain propulsion formulas can be used to deal with electromagnetic wave propagation problems. Since the FDTD method directly discretizes the time-domain wave equation, it will not limit its application range due to mathematical models, and can effectively simulate various complex structures.

The popularity of this method has strongly increased in the last two decades, mainly due to the simplicity of implementation, support of arbitrary NP shape, allowing to investigate linear and non-linear properties of NPs, using Maxwell's equations directly without approximations. There are, however, some undesired effects, like the staircase of fields in non-rectangular boundaries, mainly in code implementations without adaptive meshing. To avoid this, very fine discretization or sub-pixel smoothing of the dielectric function must be applied [56]. The dielectric function of the materials requires analytical expressions (e.g., Drude–Lorentz) [57].

FEM: Finite element method (FEM) was developed to solve differential equations of boundary-value problems [58]. Physical problems described by differential equations over a domain, like for example the Helmholtz equation in real three-dimensional space. Hence, electromagnetic (EM) field propagation around the single NP can be described by the Helmholtz Equation [59]:

$$\nabla \times \left(\mu_r^{-1}\nabla \times \boldsymbol{E}\right) - k_0^2\varepsilon_r\boldsymbol{E} = 0 \tag{6}$$

where $\boldsymbol{E}$ is the electric field of the medium, j is the current density, k_0 is the wavenumber, ε_r is the dielectric function, which is calculated as $\varepsilon_r = (n - ik)^2$, n and k are the complex refractive indices. Within each element $\boldsymbol{E}$ is approximated using a basis function expansion $\boldsymbol{E} = \sum_{j=1}^{n}\boldsymbol{N}_j\xi_j$, where the sum is over n interpolation point. $\boldsymbol{N}_j$ is chosen basis function and ξ_j is the unknown coefficient. A solution can be obtained by using the variational principle to determine ξ_j. To obtain a meaningful solution, $\boldsymbol{N}_j$ is required to satisfy Gauss's law and appropriate boundary conditions on the surface of all elements. During the last decade, an increasing number of publications on plasmonic nanostructures done with COMSOL Multiphysics has appeared in the literature.

A short comparison of these above calculation methods for NPs are listed in **Table 1**. The choice of the calculation method depends on many factors, such as: NP size, shape and dielectric environment. But the general method, such as: FEM, and FDTD, can be used in most situations by applying the periodic boundary conditions at the sides of the unit cell.

The optical properties of nanofluids also can be calculated based on the above method, such as: FDTD or FEM, which are the direct way to achieve the absorption performance of nanofluids without the strict assumptions. But the computation load is large for nanofluids since the geometry size ($\sim$ mm) is much larger than NP size or mesh size ($\sim$ nm). Therefore, the optical properties of nanofluids can be obtained from the optical properties of single NP due to the low NP concentration of plasmonic nanofluids in the solar thermal conversion applications and the independent scattering can be applied in the calculation of nanofluids.

One method is to avoid the scattering effect of NPs due to the small size, resulting in the negligible scattering effect in the nanofluids. Therefore, the absorption efficiency of the nanofluids can be obtained by the independent scattering approximation, which can be described as [9]:

$$k_{a\lambda,nf} = k_{a\lambda,bf} + k_{a\lambda,np} = \frac{4\pi\kappa}{\lambda} + \frac{f_v C_{abs}}{V_{np}} \tag{7}$$

where $k_{a\lambda,nf}$, $k_{a\lambda,bf}$, and $k_{a\lambda,np}$ are the absorption coefficients of the nanofluid, the base fluid water and the NPs respectively. κ is the absorption index of water. f_v is the NP volume fraction. V_{np} is the single NP volume. C_{abs} is the absorption cross section of the NP. Based on the Beer–Lambert law [61], the radiation intensity decays exponentially along the transmission direction. Therefore, the solar absorption efficiency η_{abs} can be calculated as:

$$\eta_{abs} = \frac{\int_{0.3\mu m}^{2.5\mu m} I_{abs}(\lambda)d\lambda}{\int_{0.3\mu m}^{2.5\mu m} I_s(\lambda)d\lambda} = \frac{\int_{0.3\mu m}^{2.5\mu m} I_s(\lambda)\left(1 - e^{-Hk_{a\lambda,nf}}\right)d\lambda}{\int_{0.3\mu m}^{2.5\mu m} I_s(\lambda)d\lambda} \tag{8}$$

where $I_s(\lambda)$ is the solar spectra at AM 1.5. $I_{abs}(\lambda)$ is the absorbed spectra. H is the depth.

Method	Description	Advantages	Limitations
Mie theory	Theoretical result	Accurate solution	Valid for simple shapes (such as: sphere)
DDA	Approximate result based on discrete dipoles	Arbitrary shape	Time-consuming; large memory space for large NPs
BEM	Numeric solution with surface discretization	Fast calculation than volume methods	Complex parameterization of boundary elements
FEM	Differential equations solved over a domain	General application	Time-consuming; large memory space for complex structures
FDTD	Discretization of Maxwell equation with Yee cell	General application	Computational stability depends on dielectric functions; Time-consuming for spectral calculations

Table 1.
Comparison of calculation methods for plasmonic NPs [60].

The optical properties of nanofluids also can be solved by the Monte Carlo (MC) method to obtain the solar absorption performance of nanofluids. MC technique is a flexible method for simulating light propagation in the medium. The simulation is based on the random walks that photons make as they travel, which are chosen by statistically sampling the probability distributions for step size and angular deflection per scattering event. After propagating many photons, the net distribution of all the photon paths yields an accurate approximation to reality. In this method, the scattering effect is considered by the scattering efficiency and scattering phase function. The absorptance, transmittance, and reflectance of nanofluids can be calculated by counting the fate of photons.

For the plasmonic nanofluids applied in the solar thermal applications, the absorption spectral distribution is one of the most important parameters, which is proportional to the NP parameters (concentration, shape and size), Qin et al. theoretically optimized the spectral absorption coefficient of an ideal plasmonic nanofluid for a DASC to maximize the thermal efficiency while maintaining the magnitude of the average absorption coefficient at a certain value [62]. However, considering that the SPR frequency of metallic NPs, such as Au, Ag, and Al, is usually located in the ultraviolet to visible range. The actual plasmonic nanofluids usually have the narrow absorption band due to the SPRs. Two strategies can be adopted to overcome this shortage in the NP theoretical design process.

One is to blend NPs with different absorption peaks to form hybrid plasmonic nanofluids for full utilization of solar energy in a broad spectrum. For example, an ideal distribution of spherical metal NPs, including nanospheres and nanoshells, were designed to match the AM 1.5 solar spectrum with an determination of absorbing and scattering distributions [63]. Based on MC method and FEM, four type of Au nanoshells were blended in the base fluid to enhance the solar absorption performance of plasmonic nanofluids with an extremely low particle concentration (e.g., approximately 70% for a 0.05% particle volume fraction) [64]. By applying the customized genetic algorithm, an optimal combination for a blended nanofluid (metal nanosphere, metal@SiO_2 core-shell, and metal nanorod) was designed with the desired spectral distribution of the absorption coefficient [65]. Besides the core-shell NPs, other NP shapes were also designed to expand the absorbance over the entire solar spectrum [66, 67]. Although different blended NPs were designed to broad the absorption spectrum, the comparison or enhancement is usually done based on the single-element nanofluid, which is not enough to compare with the other blend styles.

The other is to design complex NP structures with multiple absorption peaks at different wavelengths or coupled with the great intrinsic absorption materials. For example, core-shell NPs (Al@CdS [68], Ag@SiO_2@CdS [69], Au@C [70], Ag@TiO_2 [71], and gallium-doped zinc oxide@Cu [72, 73]) were the direct way to enhance the solar absorption performance due to the enhancement and tunable SPRs of shell and intrinsic absorption of core by optical simulations, thus broadening the absorption spectrum and improving the solar absorption performance of plasmonic nanofluids. Results also found that Ag NPs with sharp edges can induce multiple absorption peaks due to both LSPR and lightning rod effect to broad the absorption spectrum [74]. In addition, a plasmonic dimer nanofluid, consisting of the rod and sphere, was proposed to enhance the solar absorption performance by LSPR, PSPR, and gap resonance between the rod and sphere at different wavelengths [17], which was also similar as the thorny NPs [75].

3.2 Experimental design

Although various NP structures were designed to enhance the solar absorption performance of plasmonic nanofluids theoretically, the synthesizes of these complex structures are still difficult and more efforts still are needed to precious control the NP size parameters (e.g., size or shape) experimentally.

Compared with the other common nanofluids (e.g., SiO_2, TiO_2, Al), plasmonic Au nanofluids with the small NP size were prepared experimentally to obtain the great solar thermal conversion efficiency [21, 76]. However, the absorption peaks of these common metals usually locate in the visible part especially for the metal sphere. Multi-element NPs (such as: alloy NP [77]) can further enhance solar absorption ability compared with the single-element NPs by tuning the LSPR peak. Various core-shell NPs were also prepared to enhance the solar absorption performance of plasmonic nanofluids [73, 78]. For example: Ag@CdS core-shell NPs were synthesized by a facile method and the optical absorption performance of Ag@CdS nanofluids was enhanced in a wide range of visible light compared with bare Ag and CdS NPs [79]. Sn@SiO_2@Ag core-shell NPs were prepared with good abilities of both optical absorption and thermal energy storage [80]. Ag shell can improve light absorption due to LSPR effect, which was also can be found experimentally for CuO@Ag [81] and TiO_2@Ag plasmonic nanofluids compared with CuO, TiO_2, and Ag nanofluids [82].

Another simple way is to blend sphere NPs with different materials [83–85]. Various NPs have been blended experimentally to enhance the solar absorption performance of plasmonic nanofluids. For example, hybrid nanofluids containing reduced graphene oxides decorated with Ag NPs [86], multi-wall carbon nanotubes and SiO_2@Ag NPs [87], Fe_3O_4, Cu and Au NPs [88], and Au and TiN NPs [89] showed great solar absorption performance by tuning the ratios of different components to broaden the absorption spectrum. LSPR effect around plasmonic NPs and intrinsic absorption of semiconductor NPs make the hybrid nanofluids possess superior optical absorption to bare NPs at the same concentration. Besides the blended nanofluids with different NP materials discussed above, the other route is to blend the NPs with different shapes. For example, by mixing Au NPs (such as: nanorods [90]) with different shapes in water, a blended plasmonic nanofluid was prepared and absorption spectrum can be broadened due to the various LSPR peaks of different NP shapes [84]. The blended nanofluids based on Ag triangular nanosheets and Au nanorods, were proposed and a high efficiency of 76.9% is achieved experimentally with a very low volume concentration (0.0001%) [91].

As discussed above, blending different NPs is a simple way to achieve multi absorption peaks. However, compared with the single component NPs, the interaction between the different NPs in the blended nanofluids is limited due to independent scattering at the low NP fraction, leading that the solar thermal conversion efficiency of blended NPs was almost equal to the arithmetic sum of the efficiency of each component NPs without enhanced coupled effect between different NPs with the incident light [83]. Designing complex structures with multi-resonance peaks experimentally can be an efficient way to enhance the solar absorption performance of plasmonic nanofluids. For example, Au thorn [92] and Au dimer [93] were designed experimentally enhance the light absorption performance of plasmonic nanofluids. In addition, Janus NPs also showed great optical absorption performance due to the complex structure experimentally [94].

4. Applications

Nanofluids can either absorb or transmit specific solar spectrum and thus making assorted nanofluids ideal candidates for various solar applications [95]. Based on the tunable optical absorption performance of plasmonic nanofluids, several applications, including full spectrum absorption in direct solar absorption collector, selective absorption in solar PT/V systems, and local heating in solar evaporation or steam generation, are discussed below in **Figure 4**.

Some efforts have been made to investigate the solar thermal conversion performance of stationary plasmonic nanofluids based on the direct solar absorption collectors (DASCs). A one-dimensional transient heat transfer analysis was carried out to analyze the effects of NP volume fraction, collector height, irradiation time, solar flux, and NP material on the collector efficiency. Results showed that the plasmonic nanofluids (e.g., Au and Ag) achieved the better collector efficiency in the stationary state [98]. Solar thermal conversion performance of Au nanofluids in a cylindrical tube under natural solar irradiation conditions was studied and a efficiency of 76.0% at a concentration of 5.8 ppm can be achieved [99]. Although Au nanofluids have high solar absorption performance, their expensive cost limits their practical use [100]. The solar thermal conversion performance of six (Ag, Cu Zn Fe, Si and Al_2O_3) common NPs in direct absorption solar collectors (DASC) was investigated under a focused simulated solar flux. Ag nanofluid turned out to be the best among all due its strong plasmonic resonance nature [101]. Stable silver nanofluids were prepared through a high-pressure homogenizer and the outdoor experiments were conducted under sunlight on a rooftop continuously for ~10 h and the excellent photothermal conversion capability even under very low concentrations can be achieved [102].

Recently, the direct-absorption parabolic-trough solar collector (DAPTSC) using the flow nanofluids has been proposed, and its thermal efficiency has been reported to be 5–10% higher than the conventional surface-based parabolic-trough solar collector. In order to reduce the cost of a collector and avoid NP agglomeration when using plasmonic nanofluids, the configuration with the lowest possible absorption coefficient but with the reasonably high temperature gain as well as efficiency was explored [103]. For the collector design, an extra glass tube inside was inserted so the nanofluid was separated into two concentric segmentations (i.e., an inner section and an outer section), and a nanofluid of lower concentration was applied in the outer section while a nanofluid of a higher concentration in the inner section. Results showed that at the same NP concentration parameter, the DAPTSCs with two concentric segmentations of nanofluids outperform those with one uniform nanofluid for all considered configurations [104]. Furthermore, the transparent DAPTSC was improved by applying a reflective coating on the upper half of the inner glass tube outer surface such that the optical path length was doubled compared to that of the transparent DAPTSC, allowing a reduction in the absorption coefficient of the nanofluid [105]. In addition, by replacing the semi-cylindrical reflective coating with a semi-cylindrical absorbing coating for exploiting both volumetric and surface absorption of the solar radiation. The DAPTSC with a

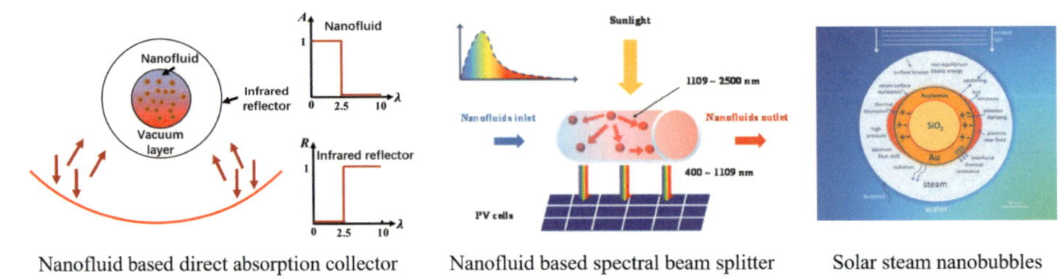

Figure 4.
Solar thermal applications of plasmonic nanofluids, including nanofluid based direct solar absorption collector, solar spectral beam splitter in solar PV/T systems [96], solar steam or nanobubble generation in solar evaporation [97].

hybrid of volumetric and surface absorption can achieve a significantly higher thermal efficiency than the previous design of a DAPTSC with a reflective coating [8]. An innovative nanofluid enabled pump-free DASC concept was presented by combining the advantages of volumetric solar harvesting and oscillating heat pipes to enhance the solar harvesting and spontaneously transfer the heat into targeted areas, providing a novel approach for efficient solar energy utilization [106].

Nanofluid-based spectral beam splitters have become dramatically popular for PV/T applications due to it can achieve tunable optical properties inexpensively [107]. For example, $CoSO_4$-based Ag nanofluid was developed to be utilized as fluid optical filter for hybrid PV/T system with silicon concentrator solar cell [108]. Furthermore, Ag NPs suspended in hybrid $CoSO_4$ and propylene glycol base fluids were prepared for both silicon and GaAs cells. Ag/$CoSO_4$-PG nanofluid filters exhibited broad absorption outside solar wavelengths and showed high transmittance in wavelength range used by the two types of cells efficiently [109]. More review about the application of nanofluids in solar PT/V systems can be found in [96, 110].

Steam generation by nanofluid under solar radiation has attracted intensive attention recently. Due to strong absorption of solar energy, NP-based solar vapor generation could have wide applications in many areas including desalination, sterilization and power generation. Steam generation of Au nanofluids under focused sunlight of 5 sun and 10 sun were performed. Results showed that localized energy trapping at the surface of nanofluid was responsible for the fast vapor generation [111]. The total efficiency reached 65% using a plasmonic Au nanofluid (178 ppm) under 10 sun, achieving a $\sim$ 300% enhancement in efficiency compared with the pure water [112]. Optimizing the range of nanofluid concentration and optical depth can be used for future solar vapor generator design. To further increase the sunlight intensity to 220 sun, experiment results coupled with the simulation model indicated that the initial stage of steam generation is mainly caused by localized boiling and vaporization in the superheated region due to highly non-uniform temperature and radiation energy distribution, albeit the bulk fluid is still subcooled [113]. A similar experiment under a sunlight intensity of 280 sun was also conducted to investigate the steam production phenomenon using Au nanofluids [114]. To further improve the solar evaporation, bubbles were also introduced into dilute plasmonic nanofluids to enhance solar water evaporation, which acted as light scattering centers to extend the incident light pathway and provided large gas–liquid interfaces for moisture capture as well as kinetic energy from bubble bursting to improve vapor diffusion [115]. Well-controlled experiments were performed to clarify the mechanism of the solar evaporation process using plasmonic Au nanofluid, carbon black nanofluid, and micro-sized porous medium. The results showed that Au nanofluids are not feasible for solar evaporation applications due to the high cost and low absorptance. High nanofluid concentration is needed to trap the solar energy in a thin layer at the liquid-gaseous interface, resulting in a local higher temperature and a higher evaporation rate [116, 117].

5. Conclusions and challenges

Plasmonic nanofluids show great interest to improve the absorption ability due to the surface plasmon resonance (SPR) around the NP surface. By designing the NP parameters (material, shape, and size) or base fluid, plasmonic nanofluids can either absorb or transmit specific solar spectrum and thus making nanofluids ideal candidates for various solar applications in full spectrum absorption in direct solar absorption collectors, selective absorption in solar PT/V systems, and local heating in solar evaporation. As discussed above, some efforts have been made to improve the solar thermal conversion applications of plasmonic nanofluids. Some challenges are still needed to overcome for the further development of plasmonic nanofluid applications. The first one is the stability of nanofluids including the long-time and high-temperature. Currently, many works were conducted in the lab and a great solar thermal conversion performance can be achieved in a short period. The performance of nanofluids in the actual applications should be considered. The second one is that the performance evaluation standard of nanofluids should be unified. Many experiments were conducted and compared in the unique experimental conditions by oneself and a more extensive evolution method is needed for different researchers to compare the performance of different nanofluids. The last one is the cost of plasmonic nanofluids. The common plasmonic metal NPs, such as: Au, Ag, Cu, are expensive in the actual applications. More cheap NPs, including the low cost of preparation processes and materials, should be developed in the further.

Acknowledgements

This work was financially supported by the Central South University and the National Natural Science Foundation of China (Grant No. 52006246).

Conflict of interest

The authors declare that they have no known competing financial interests or personal relationships that could have appeared to influence the work reported in this paper.

Author details

Meijie Chen*, Xingyu Chen and Dongling Wu
School of Energy Science and Engineering, Central South University, Changsha, China

*Address all correspondence to: chenmeijie@csu.edu.cn

References

[1] Chen M, He Y, Zhu J, Shuai Y, Jiang B, Huang Y. An experimental investigation on sunlight absorption characteristics of silver nanofluids. Sol Energy 2015;115:85–94. https://doi.org/10.1016/j.solener.2015.01.031.

[2] Wang X, He Y, Chen M, Hu Y. ZnO-Au composite hierarchical particles dispersed oil-based nanofluids for direct absorption solar collectors. Sol Energy Mater Sol Cells 2018;179:185–93. https://doi.org/10.1016/j.solmat.2017.11.012.

[3] Huang J, He Y, Chen M, Wang X. Separating photo-thermal conversion and steam generation process for evaporation enhancement using a solar absorber. Appl Energy 2019:244–52. https://doi.org/10.1016/j.apenergy.2018.11.090.

[4] Li H, He Y, Wang C, Wang X, Hu Y. Tunable thermal and electricity generation enabled by spectrally selective absorption nanoparticles for photovoltaic/thermal applications. Appl Energy 2019;236:117–26. https://doi.org/10.1016/j.apenergy.2018.11.085.

[5] Ye Q, Chen M, Cai W. Numerically investigating a wide-angle polarization-independent ultra-broadband solar selective absorber for high-efficiency solar thermal energy conversion. Sol Energy 2019;184:489–96. https://doi.org/10.1016/j.solener.2019.04.037.

[6] Chen M, He Y. Plasmonic nanostructures for broadband solar absorption based on the intrinsic absorption of metals. Sol Energy Mater Sol Cells 2018;188:156–63. https://doi.org/10.1016/j.solmat.2018.09.003.

[7] Otanicar TP, Phelan PE, Golden JS. Optical properties of liquids for direct absorption solar thermal energy systems. Sol Energy 2009;83:969–77. https://doi.org/10.1016/j.solener.2008.12.009.

[8] Qin C, Lee J, Lee BJ. A hybrid direct-absorption parabolic-trough solar collector combining both volumetric and surface absorption. Appl Therm Eng 2021;185:116333. https://doi.org/10.1016/j.applthermaleng.2020.116333.

[9] Chen M, He Y, Zhu J, Wen D. Investigating the collector efficiency of silver nanofluids based direct absorption solar collectors. Appl Energy 2016;181: 65–74. https://doi.org/10.1016/j.apenergy.2016.08.054.

[10] Choi SUS. Enhancing thermal conductivity of fluids with nanoparticles. Am. Soc. Mech. Eng. Fluids Eng. Div. FED, vol. 231, 1995, p. 99–105.

[11] Qi C, Hu J, Liu M, Guo L, Rao Z. Experimental study on thermo-hydraulic performances of CPU cooled by nanofluids. Energy Convers Manag 2017;153:557–65. https://doi.org/10.1016/j.enconman.2017.10.041.

[12] Hu Y, He Y, Gao H, Zhang Z. Forced convective heat transfer characteristics of solar salt-based SiO2 nanofluids in solar energy applications. Appl Therm Eng 2019;155:650–9. https://doi.org/10.1016/j.applthermaleng.2019.04.109.

[13] Hazra SK, Ghosh S, Nandi TK. Photo-thermal conversion characteristics of carbon black-ethylene glycol nanofluids for applications in direct absorption solar collectors. Appl Therm Eng 2019;163:114402. https://doi.org/10.1016/j.applthermaleng.2019.114402.

[14] Guo C, Liu C, Jiao S, Wang R, Rao Z. Introducing optical fi ber as internal light source into direct absorption solar collector for enhancing photo-thermal conversion performance of MWCNT- H 2 O nano fl uids. Appl Therm Eng 2020; 173:115207. https://doi.org/10.1016/j.applthermaleng.2020.115207.

[15] Zhang H, Chen HJ, Du X, Wen D. Photothermal conversion characteristics of gold nanoparticle dispersions. Sol Energy 2014;100:141–7. https://doi.org/10.1016/j.solener.2013.12.004.

[16] Prashant K. Jain, Xiaohua Huang IHE-S and MAE-S. Noble Metals on the Nanoscale: Optical and Photothermal Properties and Some Applications in Imaging, Sensing, Biology, and Medicine. Acc Chem Res 2008;41:1578–1586. https://doi.org/10.1021/ar7002804.

[17] Chen Z, Chen M, Yan H, Zhou P, Chen XY. Enhanced solar thermal conversion performance of plasmonic gold dimer nanofluids. Appl Therm Eng 2020;178:115561. https://doi.org/10.1016/j.applthermaleng.2020.115561.

[18] Li Y, Zhou J, Tung S, Schneider E, Xi S. A review on development of nanofluid preparation and characterization. Powder Technol 2009;196:89–101. https://doi.org/10.1016/j.powtec.2009.07.025.

[19] Chen HJ, Wen D. Ultrasonic-aided fabrication of gold nanofluids. Nanoscale Res Lett 2011;6:1–8. https://doi.org/10.1186/1556-276X-6-198.

[20] Daniel MC, Astruc D. Gold Nanoparticles: Assembly, Supramolecular Chemistry, Quantum-Size-Related Properties, and Applications Toward Biology, Catalysis, and Nanotechnology. Chem Rev 2004;104:293–346. https://doi.org/10.1021/cr030698+.

[21] Chen M, He Y, Huang J, Zhu J. Investigation into Au nanofluids for solar photothermal conversion. Int J Heat Mass Transf 2017;108:1894–900. https://doi.org/10.1016/j.ijheatmasstransfer.2017.01.005.

[22] He Y, Chen M, Wang X, Hu Y. Plasmonic multi-thorny Gold nanostructures for enhanced solar thermal conversion. Sol Energy 2018;171:73–82. https://doi.org/10.1016/j.solener.2018.06.071.

[23] FRENS G. Controlled Nucleation for the Regulation of the Particle Size in Monodisperse Gold Suspensions. Nat Phys Sci 1973;241:20–2. https://doi.org/10.1038/physci241020a0.

[24] Ji X, Song X, Li J, Bai Y, Yang W, Peng X. Size control of gold nanocrystals in citrate reduction: The third role of citrate. J Am Chem Soc 2007;129:13939–48. https://doi.org/10.1021/ja074447k.

[25] Bastús NG, Comenge J, Puntes V. Kinetically controlled seeded growth synthesis of citrate-stabilized gold nanoparticles of up to 200 nm: Size focusing versus ostwald ripening. Langmuir 2011;27:11098–105. https://doi.org/10.1021/la201938u.

[26] Ojea-Jiménez I, Romero FM, Bastús NG, Puntes V. Small gold nanoparticles synthesized with sodium citrate and heavy water: Insights into the reaction mechanism. J Phys Chem C 2010;114:1800–4. https://doi.org/10.1021/jp9091305.

[27] Xia Y, Xiong Y, Lim B, Skrabalak SE. Shape-controlled synthesis of metal nanocrystals: Simple chemistry meets complex physics? Angew Chemie - Int Ed 2009;48:60–103. https://doi.org/10.1002/anie.200802248.

[28] Wu HL, Kuo CH, Huang MH. Seed-mediated synthesis of gold nanocrystals with systematic shape evolution from cubic to trisoctahedral and rhombic dodecahedral structures. Langmuir 2010;26:12307–13. https://doi.org/10.1021/la1015065.

[29] Jana NR, Gearheart L, Murphy CJ. Evidence for seed-mediated nucleation in the chemical reduction of gold salts to gold nanoparticles. Chem Mater 2001;13:2313–22. https://doi.org/10.1021/cm000662n.

[30] Huang Y, Yang C, Lang J, Zhang S, Feng S, Schaefer LA, et al. Metal Nanoparticle Harvesting by Continuous Rotating Electrodeposition and Separation. Matter 2020;3:1294–307. https://doi.org/10.1016/j.matt.2020.08.019.

[31] Rechberger W, Hohenau A, Leitner A, Krenn JR, Lamprecht B, Aussenegg FR. Optical properties of two interacting gold nanoparticles. Opt Commun 2003;220:137–41. https://doi.org/10.1016/S0030-4018(03)01357-9.

[32] Atwater HA, Polman A. Plasmonics for improved photovoltaic devices. Nat Mater 2010;9:205–13. https://doi.org/10.1038/nmat2629.

[33] Gonçalves MR, Makaryan T, Enderle F, Wiedemann S, Plettl A, Marti O, et al. Plasmonic nanostructures fabricated using nanosphere-lithography, soft-lithography and plasma etching. Beilstein J Nanotechnol 2011;2:448–58. https://doi.org/10.3762/bjnano.2.49.

[34] Yu F, Chen Y, Liang X, Xu J, Lee C, Liang Q, et al. Dispersion stability of thermal nanofluids. Prog Nat Sci Mater Int 2017;27:531–42. https://doi.org/10.1016/j.pnsc.2017.08.010.

[35] Taylor RA, Phelan PE, Adrian RJ, Gunawan A, Otanicar TP. Characterization of light-induced, volumetric steam generation in nanofluids. Int J Therm Sci 2012;56:1–11. https://doi.org/10.1016/j.ijthermalsci.2012.01.012.

[36] Mahian O, Kianifar A, Kalogirou SA, Pop I, Wongwises S. A review of the applications of nanofluids in solar energy. Int J Heat Mass Transf 2013;57:582–94. https://doi.org/10.1016/j.ijheatmasstransfer.2012.10.037.

[37] Yu W, France DM, Routbort JL, Choi SUS. Review and comparison of nanofluid thermal conductivity and heat transfer enhancements. Heat Transf Eng 2008;29:432–60. https://doi.org/10.1080/01457630701850851.

[38] Wu D, Zhu H, Wang L, Liu L. Critical Issues in Nanofluids Preparation, Characterization and Thermal Conductivity. Curr Nanosci 2009;5:103–12. https://doi.org/10.2174/157341309787314548.

[39] Walshe J, Amarandei G, Ahmed H, McCormack S, Doran J. Development of poly-vinyl alcohol stabilized silver nanofluids for solar thermal applications. Sol Energy Mater Sol Cells 2019;201:110085. https://doi.org/10.1016/j.solmat.2019.110085.

[40] Lee R, Kim JB, Qin C, Lee H, Lee BJ, Jung GY. Synthesis of Therminol-based plasmonic nanofluids with core/shell nanoparticles and characterization of their absorption/scattering coefficients. Sol Energy Mater Sol Cells 2020;209:11044. https://doi.org/10.1016/j.solmat.2020.110442.

[41] Sharaf OZ, Rizk N, Joshi CP, Abi Jaoudé M, Al-Khateeb AN, Kyritsis DC, et al. Ultrastable plasmonic nanofluids in optimized direct absorption solar collectors. Energy Convers Manag 2019;199:112010. https://doi.org/10.1016/j.enconman.2019.112010.

[42] Chen Y, Quan X, Wang Z, Lee C, Wang Z, Tao P, et al. Stably dispersed high-temperature Fe3O4/silicone-oil nanofluids for direct solar thermal energy harvesting. J Mater Chem A 2016;4:17503–11. https://doi.org/10.1039/c6ta07773k.

[43] Singh N, Khullar V. Efficient Volumetric Absorption Solar Thermal Platforms Employing Thermally Stable - Solar Selective Nanofluids Engineered from Used Engine Oil. Sci Rep 2019;9:1–12. https://doi.org/10.1038/s41598-019-47126-3.

[44] Johnson PB, Christy RW. Optical constants of the noble metals. Phys Rev

B 1972;6:4370–9. https://doi.org/10.1103/PhysRevB.6.4370.

[45] David C, García De Abajo FJ. Spatial nonlocality in the optical response of metal nanoparticles. J Phys Chem C 2011;115:19470–5. https://doi.org/10.1021/jp204261u.

[46] Renger J, Quidant R, Van Hulst N, Novotny L. Surface-enhanced nonlinear four-wave mixing. Phys Rev Lett 2010; 104:046803. https://doi.org/10.1103/PhysRevLett.104.046803.

[47] Barrera RG, Fuchs R. Theory of electron energy loss in a random system of spheres. Phys Rev B 1995;52:3256–73. https://doi.org/10.1103/PhysRevB.52.3256.

[48] Link S, El-Sayed MA. Size and temperature dependence of the plasmon absorption of colloidal gold nanoparticles. J Phys Chem B 1999;103: 4212–7. https://doi.org/10.1021/jp984796o.

[49] Scholl JA, Koh AL, Dionne JA. Quantum plasmon resonances of individual metallic nanoparticles. Nature 2012;483:421–7. https://doi.org/10.1038/nature10904.

[50] Bohren CF. Absorption and scattering of light by small particles. 1983. https://doi.org/10.1088/0031-9112/35/3/025.

[51] Wiscombe WJ. Improved Mie scattering algorithms. Appl Opt 1980;19: 1505. https://doi.org/10.1364/ao.19.001505.

[52] Xu Y lin, Gustafson BÅS. A generalized multiparticle Mie-solution: Further experimental verification. J Quant Spectrosc Radiat Transf 2001;70: 395–419. https://doi.org/10.1016/S0022-4073(01)00019-X.

[53] Purcell EM, Pennypacker CR. Scattering and Absorption of Light by Nonspherical Dielectric Grains. Astrophys J 1973;186:705. https://doi.org/10.1086/152538.

[54] García De Abajo FJ. Optical excitations in electron microscopy. Rev Mod Phys 2010;82:209–75. https://doi.org/10.1103/RevModPhys.82.209.

[55] Yee KS. Numerical Solution of Initial Boundary Value Problems Involving Maxwell's Equations in Isotropic Media. IEEE Trans Antennas Propag 1966;14: 302–7. https://doi.org/10.1109/TAP.1966.1138693.

[56] Farjadpour A, Roundy D, Rodriguez A, Ibanescu M, Bermel P, Joannopoulos JD, et al. Improving accuracy by subpixel smoothing in the finite-difference time domain. Opt Lett 2006;31:2972. https://doi.org/10.1364/ol.31.002972.

[57] Hao F, Nordlander P. Efficient dielectric function for FDTD simulation of the optical properties of silver and gold nanoparticles. Chem Phys Lett 2007;446:115–8. https://doi.org/10.1016/j.cplett.2007.08.027.

[58] Salazar-Palma M, García-Castillo LE, Sarkar TK. The finite element method in electromagnetics. Eur. Congr. Comput. Methods Appl. Sci. Eng. ECCOMAS 2000, 2000.

[59] Zhao J, Pinchuk AO, Mcmahon JM, Li S, Ausman LK, Atkinson AL, et al. Methods for Describing the Electromagnetic Properties of Silver and Gold Nanoparticles. Chem Soc Rev 2008;41:1710–20.

[60] Gonçalves MR. Plasmonic nanoparticles: Fabrication, simulation and experiments. J Phys D Appl Phys 2014;47:213001. https://doi.org/10.1088/0022-3727/47/21/213001.

[61] Swinehart DF. The Beer-Lambert law. J Chem Educ 1962;39:333–5. https://doi.org/10.1021/ed039p333.

[62] Qin C, Kang K, Lee I, Lee BJ. Optimization of the spectral absorption coefficient of a plasmonic nanofluid for a direct absorption solar collector. Sol Energy 2018;169:231–6. https://doi.org/10.1016/j.solener.2018.04.056.

[63] Cole JR, Halas NJ. Optimized plasmonic nanoparticle distributions for solar spectrum harvesting. Appl Phys Lett 2006;89:153120. https://doi.org/10.1063/1.2360918.

[64] Lee BJ, Park K, Walsh T, Xu L. Radiative heat transfer analysis in plasmonic nanofluids for direct solar thermal absorption. J Sol Energy Eng Trans ASME 2012;134. https://doi.org/10.1115/1.4005756.

[65] Seo J, Qin C, Lee J, Lee BJ. Tailoring the Spectral Absorption Coefficient of aBlended Plasmonic Nanofluid Using a CustomizedGenetic Algorithm. Sci Rep 2020;10:1–10. https://doi.org/10.1038/s41598-020-65811-6.

[66] Du M, Tang GH. Plasmonic nanofluids based on gold nanorods/nanoellipsoids/nanosheets for solar energy harvesting. Sol Energy 2016;137:393–400. https://doi.org/10.1016/j.solener.2016.08.029.

[67] Mallah AR, Kazi SN, Zubir MNM, Badarudin A. Blended morphologies of plasmonic nanofluids for direct absorption applications. Appl Energy 2018;229:505–21. https://doi.org/10.1016/j.apenergy.2018.07.113.

[68] Duan H, Xuan Y. Enhanced optical absorption of the plasmonic nanoshell suspension based on the solar photocatalytic hydrogen production system. Appl Energy 2014;114:22–9. https://doi.org/10.1016/j.apenergy.2013.09.035.

[69] Duan H, Xuan Y. Enhancement of light absorption of cadmium sulfide nanoparticle at specific wave band by plasmon resonance shifts. Phys E Low-Dimensional Syst Nanostructures 2011;43:1475–80. https://doi.org/10.1016/j.physe.2011.04.010.

[70] Wang Z, Quan X, Zhang Z, Cheng P. Optical absorption of carbon-gold core-shell nanoparticles. J Quant Spectrosc Radiat Transf 2018;205:291–8. https://doi.org/10.1016/j.jqsrt.2017.08.001.

[71] Liu X, Xuan Y. Defects-assisted solar absorption of plasmonic nanoshell-based nanofluids. Sol Energy 2017;146:503–10. https://doi.org/10.1016/j.solener.2017.03.024.

[72] Yu X, Xuan Y. Solar absorption properties of embellished GZO/Cu Janus nanoparticles. Energy Procedia 2019;158:345–50. https://doi.org/10.1016/j.egypro.2019.01.100.

[73] Liu X, Xuan Y. Full-spectrum volumetric solar thermal conversion: Via photonic nanofluids. Nanoscale 2017;9:14854–60. https://doi.org/10.1039/c7nr03912c.

[74] Qin C, Kim JB, Gonome H, Lee BJ. Absorption characteristics of nanoparticles with sharp edges for a direct-absorption solar collector. Renew Energy 2020;145:21–8. https://doi.org/10.1016/j.renene.2019.05.133.

[75] Chen M, Wang X, Hu Y, He Y. Coupled plasmon resonances of Au thorn nanoparticles to enhance solar absorption performance. J Quant Spectrosc Radiat Transf 2020;250:107029. https://doi.org/10.1016/j.jqsrt.2020.107029.

[76] Chen M, He Y, Zhu J, Kim DR. Enhancement of photo-thermal conversion using gold nanofluids with different particle sizes. Energy Convers Manag 2016;112:21–30. https://doi.org/10.1016/j.enconman.2016.01.009.

[77] Chen M, He Y, Zhu J. Preparation of Au–Ag bimetallic nanoparticles for

enhanced solar photothermal conversion. Int J Heat Mass Transf 2017; 114:1098–104. https://doi.org/10.1016/j.ijheatmasstransfer.2017.07.005.

[78] Xuan Y, Duan H, Li Q. Enhancement of solar energy absorption using a plasmonic nanofluid based on TiO $_2$ /Ag composite nanoparticles. RSC Adv 2014;4:16206–13. https://doi.org/10.1039/C4RA00630E.

[79] Duan H, Xuan Y. Synthesis and optical absorption of Ag/CdS core/shell plasmonic nanostructure. Sol Energy Mater Sol Cells 2014;121:8–13. https://doi.org/10.1016/j.solmat.2013.10.011.

[80] Zeng J, Xuan Y, Duan H. Tin-silica-silver composite nanoparticles for medium-to-high temperature volumetric absorption solar collectors. Sol Energy Mater Sol Cells 2016;157: 930–6. https://doi.org/10.1016/j.solmat.2016.08.012.

[81] Yu X, Xuan Y. Investigation on thermo-optical properties of CuO/Ag plasmonic nanofluids. Sol Energy 2018; 160:200–7. https://doi.org/10.1016/j.solener.2017.12.007.

[82] Xuan Y, Duan H, Li Q. Enhancement of solar energy absorption using a plasmonic nanofluid based on TiO_2/Ag composite nanoparticles. RSC Adv 2014;4:16206–13. https://doi.org/10.1039/c4ra00630e.

[83] Chen M, He Y, Huang J, Zhu J. Synthesis and solar photo-thermal conversion of Au, Ag, and Au-Ag blended plasmonic nanoparticles. Energy Convers Manag 2016;127:293–300. https://doi.org/10.1016/j.enconman.2016.09.015.

[84] Duan H, Zheng Y, Xu C, Shang Y, Ding F. Experimental investigation on the plasmonic blended nanofluid for efficient solar absorption. Appl Therm Eng 2019;161:114192. https://doi.org/10.1016/j.applthermaleng.2019.114192.

[85] Qu J, Zhang R, Wang Z, Wang Q. Photo-thermal conversion properties of hybrid CuO-MWCNT/H_2O nanofluids for direct solar thermal energy harvest. Appl Therm Eng 2019;147:390–8. https://doi.org/10.1016/j.applthermaleng.2018.10.094.

[86] Mehrali M, Ghatkesar MK, Pecnik R. Full-spectrum volumetric solar thermal conversion via graphene/silver hybrid plasmonic nanofluids. Appl Energy 2018;224:103–15. https://doi.org/10.1016/j.apenergy.2018.04.065.

[87] Zeng J, Xuan Y. Enhanced solar thermal conversion and thermal conduction of MWCNT-SiO_2/Ag binary nanofluids. Appl Energy 2018;212:809–19. https://doi.org/10.1016/j.apenergy.2017.12.083.

[88] Jin X, Lin G, Zeiny A, Jin H, Bai L, Wen D. Solar photothermal conversion characteristics of hybrid nanofluids: An experimental and numerical study. Renew Energy 2019;141:937–49. https://doi.org/10.1016/j.renene.2019.04.016.

[89] Wang L, Zhu G, Wang M, Yu W, Zeng J, Yu X, et al. Dual plasmonic Au/TiN nanofluids for efficient solar photothermal conversion. Sol Energy 2019;184:240–8. https://doi.org/10.1016/j.solener.2019.04.013.

[90] Jeon J, Park S, Lee BJ. Analysis on the performance of a flat-plate volumetric solar collector using blended plasmonic nanofluid. Sol Energy 2016; 132:247–56. https://doi.org/10.1016/j.solener.2016.03.022.

[91] Wang TM, Tang GH, Du M. Photothermal conversion enhancement of triangular nanosheets for solar energy harvest. Appl Therm Eng 2020;173: 115182. https://doi.org/10.1016/j.applthermaleng.2020.115182.

[92] Chen M, He Y, Ye Q, Wang X, Hu Y. Shape-dependent solar thermal conversion properties of plasmonic Au

nanoparticles under di ff erent light fi lter conditions. Sol Energy 2019;182: 340–7. https://doi.org/10.1016/j.solener.2019.02.070.

[93] Huang J, Liu C, Zhu Y, Masala S, Alarousu E, Han Y, et al. Harnessing structural darkness in the visible and infrared wavelengths for a new source of light. Nat Nanotechnol 2016;11:60–6. https://doi.org/10.1038/nnano.2015.228.

[94] Zeng J, Xuan Y. Tunable full-spectrum photo-thermal conversion features of magnetic-plasmonic Fe_3O_4/TiN nanofluid. Nano Energy 2018;51: 754–63. https://doi.org/10.1016/j.nanoen.2018.07.034.

[95] Sajid MU, Bicer Y. Nanofluids as solar spectrum splitters: A critical review. Sol Energy 2020;207:974–1001. https://doi.org/10.1016/j.solener.2020.07.009.

[96] Liang H, Wang F, Yang L, Cheng Z, Shuai Y. Progress in full spectrum solar energy utilization by spectral beam splitting hybrid PV / T system. Renew Sustain Energy Rev 2021;141:110785. https://doi.org/10.1016/j.rser.2021.110785.

[97] Polman A. Solar steam nanobubbles. ACS Nano 2013, 7(1): 15–18. https://doi.org/10.1021/nn305869y.

[98] Chen M, He Y, Zhu J, Wen D. Investigating the collector efficiency of silver nanofluids based direct absorption solar collectors. Appl Energy 2016;181: 65–74. https://doi.org/10.1016/j.apenergy.2016.08.054.

[99] Jin H, Lin G, Bai L, Amjad M, Bandarra Filho EP, Wen D. Photothermal conversion efficiency of nanofluids: An experimental and numerical study. Sol Energy 2016;139: 278–89. https://doi.org/10.1016/j.solener.2016.09.021.

[100] Zeiny A, Jin H, Bai L, Lin G, Wen D. A comparative study of direct absorption nanofluids for solar thermal applications. Sol Energy 2018;161:74–82. https://doi.org/10.1016/j.solener.2017.12.037.

[101] Amjad M, Jin H, Du X, Wen D. Experimental photothermal performance of nanofluids under concentrated solar flux. Sol Energy Mater Sol Cells 2018;182:255–62. https://doi.org/10.1016/j.solmat.2018.03.044.

[102] Bandarra Filho EP, Mendoza OSH, Beicker CLL, Menezes A, Wen D. Experimental investigation of a silver nanoparticle-based direct absorption solar thermal system. Energy Convers Manag 2014;84:261–7. https://doi.org/10.1016/j.enconman.2014.04.009.

[103] Qin C, Kang K, Lee I, Lee BJ. Optimization of a direct absorption solar collector with blended plasmonic nanofluids. Sol Energy 2017;150:512–20. https://doi.org/10.1016/j.solener.2017.05.007.

[104] Qin C, Kim JB, Lee J, Lee BJ. Comparative analysis of direct-absorption parabolic-trough solar collectors considering concentric nanofluid segmentation. Int J Energy Res 2020;44:4015–4025. https://doi.org/10.1002/er.5165.

[105] Qin C, Kim JB, Lee BJ. Performance analysis of a direct-absorption parabolic-trough solar collector using plasmonic nanofluids. Renew Energy 2019;143:24–33. https://doi.org/10.1016/j.renene.2019.04.146.

[106] Jin H, Lin G, Guo Y, Bai L, Wen D. Nanoparticles enabled pump-free direct absorption solar collectors. Renew Energy 2020;145:2337–44. https://doi.org/10.1016/j.renene.2019.07.108.

[107] Du M, Tang GH, Wang TM. Exergy analysis of a hybrid PV/T system based on plasmonic nanofluids and silica aerogel glazing. Sol Energy 2019;183: 501–11. https://doi.org/10.1016/j.solener.2019.03.057.

[108] Han X, Chen X, Wang Q, Alelyani SM, Qu J. Investigation of $CoSO_4$-based Ag nanofluids as spectral beam splitters for hybrid PV/T applications. Sol Energy 2019;177:387–94. https://doi.org/10.1016/j.solener.2018.11.037.

[109] Han X, Chen X, Sun Y, Qu J. Performance improvement of a PV/T system utilizing Ag/$CoSO_4$-propylene glycol nanofluid optical filter. Energy 2020;192:116611. https://doi.org/10.1016/j.energy.2019.116611.

[110] Abdelrazik AS, Al-Sulaiman FA, Saidur R, Ben-Mansour R. A review on recent development for the design and packaging of hybrid photovoltaic/thermal (PV/T) solar systems. Renew Sustain Energy Rev 2018;95:110–29. https://doi.org/10.1016/j.rser.2018.07.013.

[111] Jin H, Lin G, Zeiny A, Bai L, Wen D. Nanoparticle-based solar vapor generation: An experimental and numerical study. Energy 2019;178:447–59. https://doi.org/10.1016/j.energy.2019.04.085.

[112] Wang X, He Y, Liu X, Shi L, Zhu J. Investigation of photothermal heating enabled by plasmonic nanofluids for direct solar steam generation. Sol Energy 2017;157:35–46. https://doi.org/10.1016/j.solener.2017.08.015.

[113] Jin H, Lin G, Bai L, Zeiny A, Wen D. Steam generation in a nanoparticle-based solar receiver. Nano Energy 2016;28:397–406. https://doi.org/10.1016/j.nanoen.2016.08.011.

[114] Amjad M, Raza G, Xin Y, Pervaiz S, Xu J, Du X, et al. Volumetric solar heating and steam generation via gold nanofluids. Appl Energy 2017;206:393–400. https://doi.org/10.1016/j.apenergy.2017.08.144.

[115] Yao G, Xu J, Liu G. Solar steam generation enabled by bubbly flow nanofluids. Sol Energy Mater Sol Cells 2020;206:110292. https://doi.org/10.1016/j.solmat.2019.110292.

[116] Zeiny A, Wen D. Nanofluids-based and porous media-based solar evaporation: A comparative study. AIP Conf. Proc., vol. 2123, 2019, p. 020088. https://doi.org/10.1063/1.5117015.

[117] Zeiny A, Jin H, Lin G, Song P, Wen D. Solar evaporation via nanofluids: A comparative study. Renew Energy 2018;122:443–54. https://doi.org/10.1016/j.renene.2018.01.043.

Chapter 8

Nanocomposite and Nanofluids: Towards a Sustainable Carbon Capture, Utilization, and Storage

Ronald Nguele, Katia Nchimi Nono and Kyuro Sasaki

Abstract

Large volumes of unconventional fossil resource are untapped because of the capillary forces, which kept the oil stranded underground. Furthermore, with the increasing demand for sustainable energy and the rising attention geared towards environment protection, there is a vital need to develop materials that bridge the gap between the fossil and renewable resources effectively. An intensive attention has been given to nanomaterials, which from their native features could increase either the energy storage or improve the recovery of fossil energy. The present chapter, therefore, presents the recent advancements of nanotechnology towards the production of unconventional resources and renewable energy. The chapter focuses primarily on nanomaterials applications for both fossils and renewable energies. The chapter is not intended to be an exhaustive representation of nanomaterials, rather it aims at broadening the knowledge on functional nanomaterials for possible engineering applications.

Keywords: nanoparticle, nanocomposite, oil recovery, CO_2 sequestration, solar energy

1. Introduction

Metal Oxides (MO) are a class of compounds that are as abundant in the nature than in the library of synthetic inorganic compounds. While the use of MO as bulk materials is widely applied, developing new class of materials based on MO, understanding their chemistry, tuning their characteristics, and developing novel potential engineering are still under investigation. MO nanoparticles (MO-NP) are MO with a diameter ranging between 10 and 100 nm with a high surface-to-volume ratio, which explains their advantageous features compared to similar materials of micro- or macro-size [1].

MO-NPs can contain either a single metallic specie (nanoparticle, NP) or two or more different metallic species (nanocomposite, NCP). When the metal species are separated and independent at the molecular level, the nanocomposites could refer to a simple mixture of oxides with several component phases. In some systems, nanocomposites consist in stoichiometric (or non-stoichiometric) replacement of a metal ion by another one in the first oxide crystallographic structure. Such substitution systems are referred as mixed MOs e.g. spinels [2] or perovskites [3, 4].

MO-NPs also include core-shell architectures where each particle has an inner core made of one type of MO, and the outer shell consisting in another MO. Just as single nanoparticles, composite nanoparticles exhibit size confinement properties (optical and electronic). Their composition and the atomic order of aggregates are pivotal factors for their specific features, which are tailored during their synthesis. MO-NPs have a high surface-to-volume ratio, which increases the reactivity. Dispersed in base fluid (BF), MO-NP enhances the native properties of the solution compared to the case in which no MO-NPs were added.

This chapter does not intend to list, in an exhaustive manner, the features of nanofluid. Rather, the authors aim to discuss, from both the chemistry and the engineering point of view, the key features of MO-NFs transferable to the carbon capture utilization and storage (CCUS). The chapter will cover the different synthesis methods of the NP and NCP, the formulation of NF as well as the application in respect of CCUS.

2. Synthesis of MO-NPS and MO-NCPs

MO-NPs can be obtained through two opposite approaches including top-down and bottom-up. The former technique consists in successive mechanical operations of divisions and fragmentations, or in the irradiation of the bulk phase with a powerful energy source including UV, X-Rays, electron beam [5]. Using the bottom-up approach, MO-NPs are formed within the reaction medium, subsequently to the clustering of single atoms. The growth in size is time-dependent, and may be facilitated by additional treatment such as calcination [6].

2.1 Physical methods

2.1.1 Spray pyrolysis

Spray pyrolysis is generally used for the preparation of thin-films or pulverulent materials. It consists in the spraying of a solution or a suspension containing the metallic precursors in an oven, followed by a high temperature treatment. This method allows the formation of spherical oxides as the shape of the oxides are strongly dependent of the drops generated at the entrance of the furnace. Given the rapid rate of nucleation at high temperature, there is a one-droplet, one-particle mechanism.

In addition to the shrinkage occurring following the formation of oxides, micrometers precursors can allow the formation of particles in the nanometer size range [7]. NCPs can also be obtained through this method by mixing several metal ions in the precursor solution. The formulation of the composites is controlled by adjusting the stoichiometric proportions of each metallic species in the solution [8, 9].

2.1.2 Chemical vapor deposition (CVD)

This technique is used mostly for the preparation of 2D metallic or inorganic materials of nanometric thickness. Usually, volatiles precursors are delivered on a heated surface on which a thin layer of materials is deposited upon a chemical reaction in vapor phase. Additional physical processes such as evaporation or sputtering are usually required to complete the synthesis.

During the preparation of MOs by CVD, oxidation and hydrolysis are the primary chemical reactions taking place in the presence of oxidizing agents (oxygen or ozone) and the precursors are usually metal alkoxides [10].

2.1.3 Sonochemistry

Sonochemistry is a branch of chemistry supported by the formation, growth, and collapse of bubbles in a liquid upon irradiation with high intensity ultrasound waves. The bubbles can reach a temperature and pressure as high as 5000°C and 500 mPa respectively [11]. These conditions increase the chemical reactivity of the species in the reactor. When the water is the solvent, radical •OH, H_2O_2 and O_3 are generated, leading thereby to oxidant medium suitable for the preparation of MO.

Treatment of solutions of copper (Cu), zinc (Zn) and cobalt (Co) acetates under a high-intensity ultrasonic horn has been used to produce nanosized CuO, ZnO, and CoO_3 respectively [12]. This method has also been used for the preparation of nanocomposite when the suitable precursors are mixed [13].

2.2 Chemical methods

2.2.1 Hydrothermal synthesis

Hydrothermal or solvothermal synthesis (if water is the solvent) is a synthesis method in which the precursors (dissolved or dispersed in water) are placed in an autoclave where the reaction takes place at high temperature and pressure [14]. Hydrothermal allows the synthesis of MO-NPs from a wide variety in shape, size, structure, and composition, provided that key-factors such as temperature, pressure, pH, concentration of reactants are well-controlled [15–20].

2.2.2 Sol-gel synthesis

It involves the hydrolysis and condensation of metal alkoxides, acetates, nitrates, sulfates, and chlorides. Their hydrolysis in solution results from the dispersion of metal hydroxides, which further undergo condensation leading to the formation of 3-dimensional network namely a gel. The gel is either dried by removing the solvent or treated by chemical reaction to give the condensed MO materials [21]. The solvent used is generally water (aqueous sol-gel). In most non-aqueous synthesis, additives such as surfactant can be required to control the morphology of the particles and most importantly to prevent/reduce their aggregation [22, 23].

2.2.3 Co-precipitation

The co-precipitation method entails the co-precipitation of cations in an aqueous media by the addition of an alkaline solution. A short burst of nucleation occurs when the concentration of the species reaches its critical super-saturation, and then, there is a slow growth of the nuclei by diffusion of the solutes to the surface of the crystal. MO-NPs are dried further or sometimes calcined at temperatures above 500°C, in which case there is an alteration of the structure of material [24].

The co-precipitation method can be relevant for the synthesis of both MO-NP and MO-NCP [25]. The size and morphology of the nanoparticles produced by this method can be controlled by varying the molar ratio of the two precursor ions, the base used, mixing rate, the pH, and the temperature [26]. Unfortunately, it produces wastewaters with very basic pH needs to be treated before been discarded.

2.3 Alternative synthesis methods

Most of the methods of synthesis of nanomaterials allow a great flexibility in the choice of the precursors and the morphology of the material through the control of key parameters such as temperature, concentration, or pH. However, they also required the use of toxic chemical reactants or additives. In this regard, alternative synthesis routes have been exploited, with mitigated results, including biosynthesis, plasma assisted. Extraction from ores as for it fits in the strategy of valuation of naturally occurring ores.

2.3.1 Biosynthesis of MOs NPs

This synthetic approach includes the use of vegetal substances (plant extracts or agricultural wastes) but can also involve living organisms such as bacteria and fungi. Natural extracts used for the preparation of the MO materials are available and renewable, in addition to contain organic groups that can react with the metal ions is solution in the absence of further additives [27, 28].

2.3.2 Plasma assisted synthesis

The plasma state is obtained from the ionization of noble gases, which increases the reactivity of the medium rich in electrons and chemical species upon reaching a very high temperature, up to 10000K [29]. Non-thermal plasma is an advantageous alternative that can be operated at atmospheric pressure [30]. Among the different types of systems used for plasma production, gliding arc discharge is reported to attractive for MO-NPs synthesis.

Under humid air as feeding gas, glidarc discharge produces acidic and very oxidizing environment in which many transitions metal ions precipitate as hydroxides or oxides [31–33]. However, one of the major disadvantages associated to this technique is the poor control on the size range and morphology of the particles.

2.3.3 Extraction from ores

Mineral ores are naturally occurring rock, from which the metal content can be extracted via hydrometallurgy [34]. During MO extraction, the ore is successively crushed, roasted, leached, and precipitated. The leaching step consists in the solubilization of the metal species under its ionic forms by the action of a strong acid or a strong base. Usually, impurities remain insoluble and are removed from the mixture.

The metal ions remaining in solution are then concentrated by solvent (water) evaporation before further treatment for the precipitation. When the mineral contains one major species, single MO particles are produced, but for more complex ore, composite materials can be obtained [35, 36].

3. Preparation and stability of metal oxide nanofluid

3.1 Preparation

Preparation of NFs is the most challenging step, as far as the experimental studies with NFs are concerned. This is so because nanofluids need special requirements including stable suspension, low agglomeration of particles, and no chemical change of the fluid.

3.1.1 Preparation using direct method or one-step approach

One-step technique (or direct approach) consists in formulating the nanofluid right after their synthesis. The literature reports two major approaches including direct evaporation and the laser ablation method. In either approach, the nanofluid is obtained after a transition from the gas to solid phase [37, 38].

3.1.2 Preparation using two-step method approach

Two step approach is designed in a manner that the surface of MO-NPs is sufficiently wet so to mitigate the particle impingement [39]. This is done by either using mechanical mixing [40] or sonication or the combination of both [41, 42] as shown in **Figure 1**.

During the mechanical mixing, the large flocs are broken down and the individual particles are kept separated from each other. During the sonication stage, the ultrasound waves, by stretching the molecular spacing of fluid, creates cavities within, which cause subsequently the chemical bonds to break [44]. However, due to the intrinsic properties of MO-NPs, a poor dispersion is often yielded which is mitigated by either extending the mixing time, adding few drops of acidifying agents [45] or even bubbling gas during the preparation [43].

3.2 Stability of MO-NFs

3.2.1 Monitoring the stability of NFs

3.2.1.1 Sedimentation and centrifugation method

Sedimentation method is the simplest and most straightforward method to investigate the stability of a nanofluid. A fixed volume of nanofluid is transferred into a graduated test tube and observed over the time. **Figure 2** shows the sedimentation of alumina-based NFs upon increasing the load in NPs.

The monitoring of the volume of deposited NPs showed that a load in alumina-NP of 0.05 wt.% was sufficient to give a stable nanofluid. The literature reports similar results, in which no or little visual sedimentation of particles can be observed from the naked eye [46].

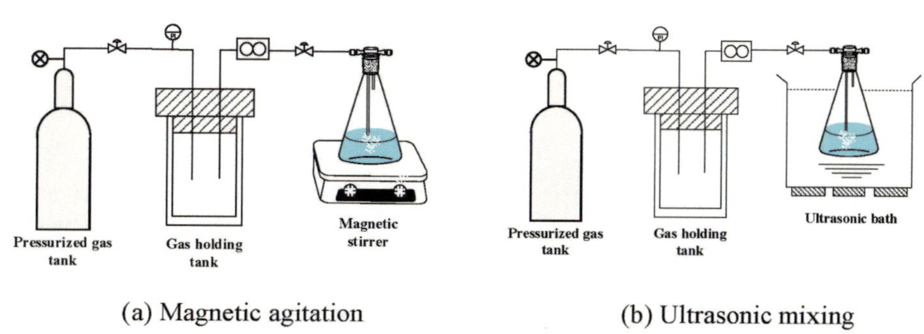

Figure 1.
Preparation of polymer-based NFs; Adapted with permission from [43]. Copyright (2019) American Chemical Society.

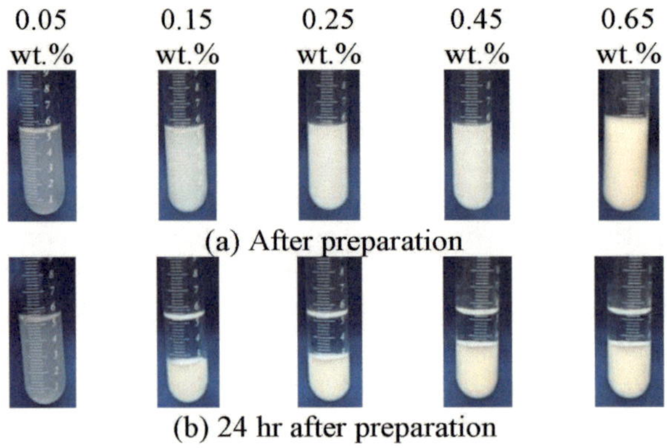

Figure 2.
Sample pictures of water-based NFs prepared using alumia oxide NPs and ethylene glycol as BF.

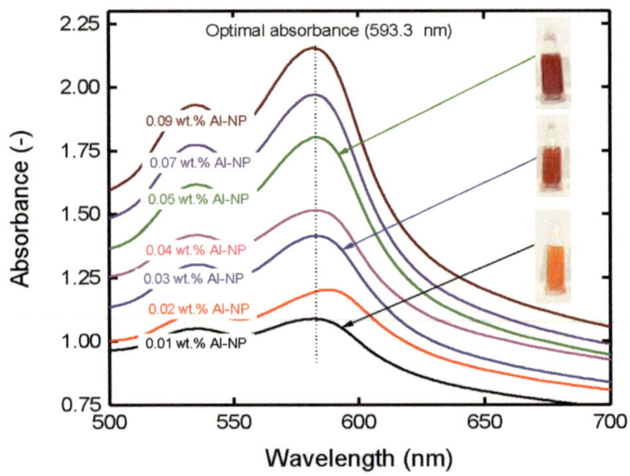

Figure 3.
Monitoring alumina-based NFs stability using UV-Vis spectroscopy and 1-(-2pyridylazo)-2-naphthol, PAN. Adapted with permission from [47]. Copyright (2020) American Chemical Society.

3.2.1.2 Spectrophotometric analyses

This approach relies on the intensity of absorption when the light passes through a target sample. As shown in **Figure 3**, Ngo et al. monitored the stability of alumina-based nanofluid by combining colorimetry and spectrophotometry [47].

3.2.1.3 Other monitoring methods

Another straightforward approach for monitoring the nanofluid stability is to measure the particle size at different time intervals. This could be achieved by either using scanning/transmitting electron micro- scope (SEM/TEM) or zeta potential [48]. SEM/TEM allows to directly visualize the distribution of particle size and the evolution of particle coagulation. Easy and fast, SEM/TEM does not require separation of NP from the solvent [49].

As far as colloidal suspensions are concerned, Zeta potential defines the electro kinetic potential in a nanofluid. It indicates the interaction energy between particles. High absolute value of zeta potential means stronger repulsive force between NPs, and hence indicates better stability of nanofluid.

3.2.2 Enhancing the stability of nanofluid

The stability of NF is fully acquired when there is a minimization of the surface area of the NPs dispersed in the solution. To prevent the particle agglomeration, an energy barrier must be created to prevent them from passing the unstable to stable energy state. This could be performed by (a) by addition of acidic or alkaline materials, (b) altering the preparation step, and (c) choosing a proper BF [50–52]. **Figure 4** shows the influence of the preparation step of nanofluid stability.

The results showed that the acidity of the solution decreases regardless the preparation method. However, combing both the sonication and the magnetic stirring could prolong the stability of the nanofluid. Furthermore, Nguele et al. [43] and later Ngo et al. [47] reported that bubbling gas during the preparation could further enhance the stability regardless the type of base fluid (**Figure 5**).

The average decrease in acidity of about 20 % from the initial value (pH =5.4) was observed throughout the preparation stage when CO_2 gas, which contrasts with an increase in pH twice higher when O_2 was bubbled. Regardless the reason pertaining to the increase in pH (i.e., carbonation for CO_2 bubbling and radical formation for O_2 bubbling), the surface modification of NP and thus the stability is enhanced.

The addition of dispersants is an alternative for enhancing the stability of NFs [48, 50, 53]. These dispersants attach to the surface of the NP due to the mutual affinity. In addition, the tail of the attached dispersant works as a steric barrier, which prevents the particles from agglomerating. Such effect, known as steric hindrance, inhibits the coagulation of NPs in the suspensions (**Figure 6**).

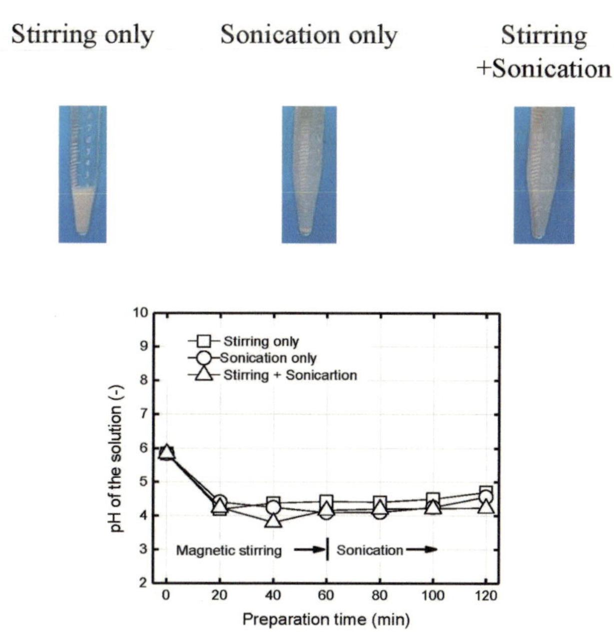

Figure 4.
Influence of preparation step on the stability of nanofluid; the nanofluid consists in Si-NP dispersed in an aqueous polymeric solution, Reprint with permission from [43] Copyright (2019) American Chemical Society.

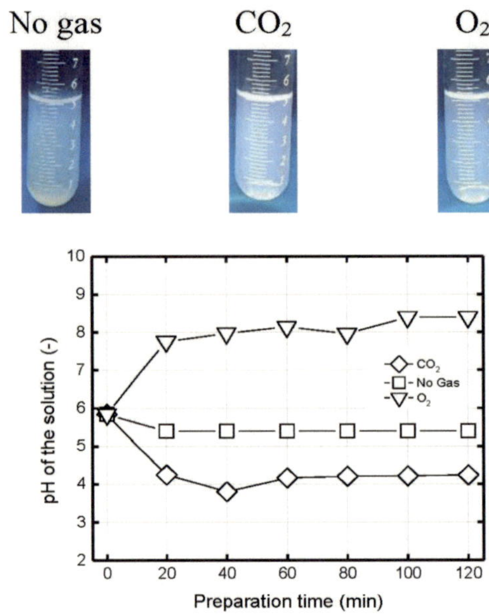

Figure 5.
Influence of gas bubbling on the stability of nanofluid; the nanofluid consists in Si-NP dispersed in a deionized water.

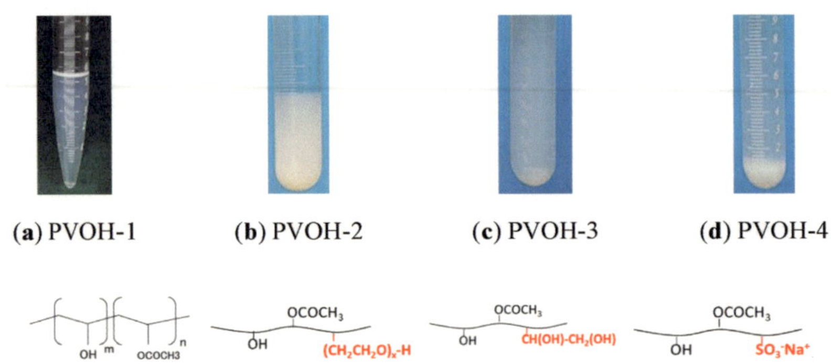

Figure 6.
Influence of BF on the stability of nanofluid at 25°C; the nanofluid consists in Si-NP dispersed in a polymer (PVOH) solution prepared following two-step approach using CO_2 bubbling.

4. Application of MO-NFs to fossil, gas storage and renewable energy

4.1 Fossil energy and enhanced oil recovery (EOR)

4.1.1 Tertiary oil recovery

Improved oil recovery (IOR) or enhanced oil recovery (EOR) are two terms used loosely to describe the improvement of oil recovery after both the primary and the secondary stages of oil production become economically unattractive or technically not feasible. In principle, IOR is the general term to designate any implemented means after secondary process that increases considerably the amount of oil recovered. On the hand, EOR defines a specific technique (or a combination of techniques) implemented to decrease the residual oil.

EOR methods are grouped into thermal and non-thermal methods. Thermal methods are the most advanced techniques among EOR methods and are best suited for heavy oils and tar sand formations. In these methods, the heat is supplied to the reservoir in form of steam or fire, which favors the vaporization of stranded oil. The major drawbacks associated to thermal-EOR pertain to the geometry and the petro-physical properties of the candidate formation [54].

Non-thermal methods encompass techniques that reduce the interfacial tension (IFT) between the stranded oil and resident fluids and the viscosity of the oil. Among the most prominent methods, gas-EOR stands out because of it offers the possibility to sequester greenhouse gases. During a gas-EOR, the injected gas dissolves into the oil after a first (or multiple) contact leading to a foamy oil, whose viscosity is lower than that of the original oil [55]. Often, gas-EOR is challenged, in part by, the deposition of heavy fractions [56–58].

If the slug is a surfactant [59], a polymer [60], or even a micellar solution [61], the oil is produced by reduction of IFT or wettability. Major problems associated to surfactant-EOR are the loss of chemical, phase partitioning and trapping, and the slug by passing. Unlike chemical-EOR, microemulsion-EOR relies on the reduction of the mobility ratio. Microemulsions, kinetically more stable than emulsions, are potentially viable because of the ultra-low IFT and their high interfacial area [62–64].

Microbial-EOR uses the potential of microbes to yield either bio-surfactant, slimes (polymers), biomass and/or gases such as CH_4, CO_2, N_2 and H_2 as well as solvents and certain organic acids [65, 66]. Oil recovery mechanisms in microbial-EOR are like those of the classic chemical methods. This includes IFT reduction, emulsification, wettability alteration, improved mobility ratio, selective plugging, viscosity reduction, oil swelling and increased reservoir pressure due to the formation of gases [67, 68].

Nano-EOR has been reported to be the next generation of EOR, which is evidenced by the wealth in literature covering the topic [69–75]. Therefrom, it appears that the mechanisms of oil production using MO-NPs are (1) wettability and the IFT alteration, (2) advanced drag reduction, and (3) the decrease of the mobility ratio.

4.1.2 EOR mechanisms in respect of MO-NPs

4.1.2.1 Wettability and IFT alteration

Wettability is the preferential tendency of one fluid to wet (or to spread) onto a surface [76]. To produce more oil, the wettability of the oil-water-rock system should be shifted from oil-wet to a water-wet or strongly water wet condition. MO-NPs can adsorb onto the rock surface and form a nanotexture, which contributes to wettability alteration [77]. However, these mechanisms are affected by the formation salinity (**Figure 7**).

At a low salt concentration, the activity coefficient of the salt increases in a manner that the salt molecules sit within the oil phase. With the presence of salt at the interface, the excess surface concentration turns positive from which results a low contact angle (**Figure 7a**) and higher IFT (**Figure 7b**). An oil production scenario in which the salt concentration is large, the salting-out effect seems to prevail [47].

MO-NPs are depleted at the interface and transferred back to oil phase. This breaks the oil-water interface adsorption, hence a high contact angle. The same behavior could be extended when two immiscible liquids (oil and water) meet each

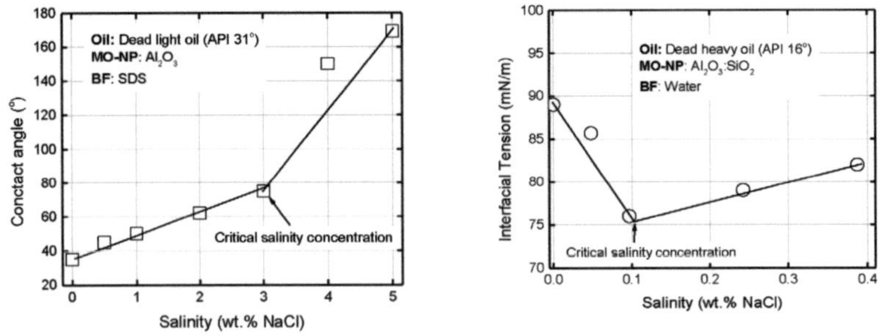

(a) Wettability alteration modified from adapted with permission from [47]. Copyright (2020) American Chemical Society

(b) IFT alteration; this study

Figure 7.
Influence formation salinity on wettability and IFT alteration. (a) Wettability alteration modified from adapted with permission from [47]. Copyright (2020) American Chemical Society. (b) IFT alteration; this study.

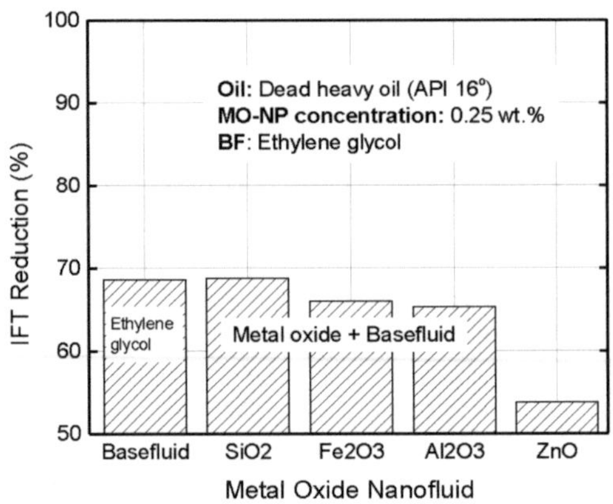

Figure 8.
Influence of type of MO-NPs on IFT alteration.

other. The molecules at the surface of both of those liquids become unbalanced forces of attraction, which cause the IFT to rise.

Adding MO-NPs could not only reduce the IFT but also the contact angle. For example, it was found adding only 0.25 wt.% of MO-NP to a polymeric BF, the contact angle as well as the IFT between the nanofluid and heavy oil (API 16°) decreases about 50% from its initial value (**Figure 8**).

4.1.2.2 Improve mobility ratio of injected fluids

The mobility ratio of water to oil is one of the most critical factors to influence water flood efficiency. When mobility is greater than one, it is considered unfavorable as the displacing fluid is more mobile than oil in the porous medium; the slug tends to bypass oil and early breakthrough is experienced at the producers (channeling). At a mobility ratio of less than one, water is less mobile than oil leading to better displacement and recovery of oil.

The mobility ratio can be decreased either by reduction of the viscosity of the resident oil or by increasing that of the nanofluid. As shown in **Figure 9**, increasing the load in MO-NP (SiO_2-NP in this experiment) prompted an increase in oil recovery in a waterflooded sandstone.

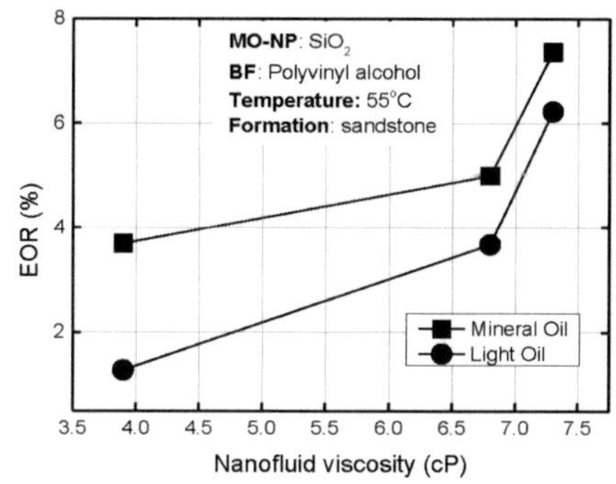

Figure 9.
Relationship between MO-NP viscosity and oil recovery factor.

The experiments were conducted using light mineral (specific density 0.838, viscosity of 26 cP at 25°C) and light crude oil (specific density 0.860 viscosity of 9.54 cP at 25°C). It was found a higher production when light mineral oil was used. This is because of the difference in native composition including a low acid number, a high concentration of asphaltene.

4.1.2.3 Pore channels plugging

Pore channels plugging can be caused by two mechanisms: mechanical entrapment and log-jamming. These mechanisms were evaluated in this study by the injection of Si-NP dispersed in aqueous polymeric solution. Two types of formations were considered including a homogeneous formation with a uniform porosity and a heterogeneous formation with contrasted porosity. The results are shown in **Figure 10**.

The production in homogeneous formation decreases monotonically with the load in Si-NP, while a reverse trend was observed for a heterogeneous model. In a homogeneous formation, the increase in MO-NP load causes the plugging of pore throats, whose size are smaller than the average size in MO-NP dispersed (log jamming). As the nanofluid travels within the formation, the narrowing of flow area and the differential pressure led to a velocity increase of the nanofluid.

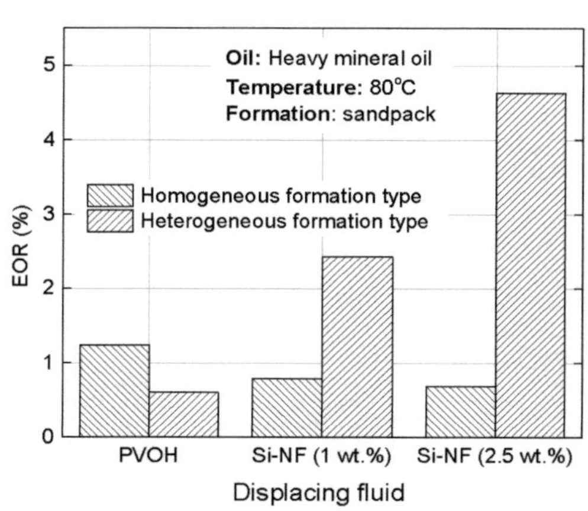

Figure 10.
Oil recovery using Si-NPs in contrasted sandstone formations.

The smaller molecules will flow faster than causing accumulation of MO-NP at the entrance of the pore throats. For larger load, there is a possibility of having a plugging at the entrance of the throat due to the size of the nanofluid (mechanical entrapment). For formations with contrasted porosity, the log-jamming or mechanical entrapment could be beneficial.

As the pore is plugged, there is a pressure build in the adjacent pore throat, forcing out the oil trapped in the pore throat or the water to move to a layer with lower porosity. This can be considered as temporary log-jamming. This phenomenon is mainly governed by the concentration and size of NPs, flow rate and the diameters of pore throats (**Figure 11**).

4.1.2.4 Preventing asphaltene precipitation

Asphaltene precipitation can cause severe problems due to the deposition inside the reservoir, at the wellhead, and/or inside the pipelines. However, it is believed MO-NPs have the potential to inhibit the adsorption and thus delay the deposition [78, 79]. The particles, in contact with the asphaltenes molecules can minimize the interactions asphaltene-asphaltene and/or asphaltene-rock leading therefore to a mitigation (**Figure 12**).

In this regard, MO-NPs are suitable candidates because their inherent properties. In this study, asphaltenes were extracted from dead heavy crude oil (API 16°) as per the procedure discussed by Goual [80]. An asphaltenic solution of 1wt.% was prepared by diluting extracted asphaltenes with toluene. Two set of experiments were conducted at room temperature including porosity impairment (**Figure 13a**) and adsorption on sandstone (**Figure 13b**).

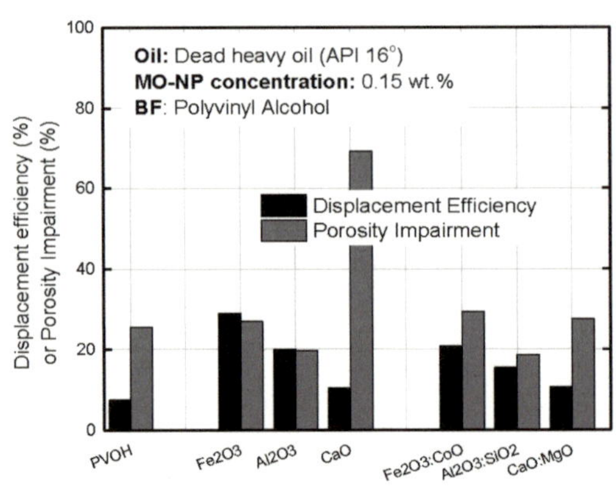

Figure 11.
Relationship between the displacement efficiency, porosity impairment and type of MO-NPs.

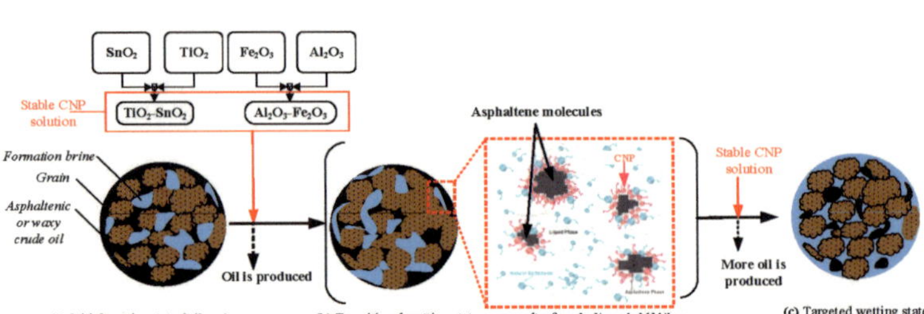

Figure 12.
Conceptual approach of asphaltene inhibition during CO_2 injection.

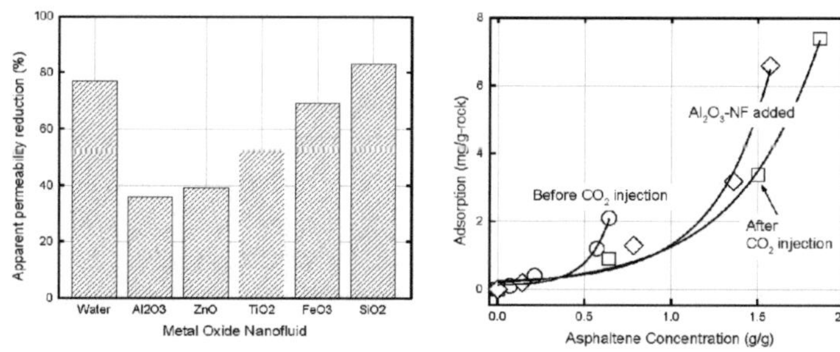

(a) (b)

Figure 13.
Asphaltene mitigation by addition of different types of MO-NPs dispersed into water. (a) Porosity impairment after CO_2 injection. (b) Static adsorption after CO_2 injection.

It could be seen that the porosity of the waterflooded sandstone decreases upon injection CO_2 (**Figure 13a**). Adding MO-NPs to the same water, the impairment could be improved with lowest obtained for Al-NP. This is so because of the higher adsorption capacity of MO-NPs, which interacts more strongly with the asphaltenes. The influence of MO-NP is noticeable as evidenced by the decrease in adsorption (**Figure 13b**).

4.2 Application of MO-NFs to CO_2 Sequestration

As of 2018, 70% of the global warming was subsequent to the release of greenhouse gases (GHG) to the atmosphere, with fossil resources contributing to up to 37.1 billion metric tons. The total concentration in carbon dioxide, CO_2, in the atmosphere was reported to hit its highest level ever (407 ppm) million. Great efforts should be invested to reduce CO_2 concentration to an acceptable value. Carbon dioxide capture, utilization and storage (CCUS) technology of which Carbon capture and storage (CCS) technologies have a potential to reduce CO_2 emissions to the atmosphere due to the huge global capacity for underground storage [81]. With 21 large-scale CCS projects operating worldwide, the volume of storable CO_2 is estimated to be up to 37 Mtpa. Yet more CCS projects are needed to reach the Paris 2 °C target, which is partly due to the leakage of the stored CO_2 through the faults of the formation within which the gas is trapped [82, 83]. A typical CCS project encompasses the capture, the compression and transport, and the injection in the designed formation. The success of a gas storage depends primarily on the trapping mechanisms occurring during CO_2 containment.

A trapping mechanism refers a process (either physical or chemical), which improves the sequestration of CO_2. Among the different known trapping mechanisms, three processes stand out including residual, solubility and mineral trapping [84]. During the residual trapping, CO_2, injected at its supercritical state, displaces the fluids as it moves through the porous rock. As CO_2 moves upward due to the buoyancy difference, some of the CO_2 will be left behind as disconnected droplets within the pore throats, which are immobile.

This mechanism, however, is challenged by the faults present the geological formation (cap rock). The fault could crack due to the over-pressurization of the aquifer leading to a leak in CO_2 Solubility trapping involves the dissolution of supercritical CO_2 in the salty water (brine), which leads to a fluid denser than the native fluids. From the difference in buoyancy, the resulting fluids force CO_2 to sink at the bottom of formation over time. The problem, in here, is that not only the solubility of CO_2 in brine is low, but it reaches quickly its saturation causing thereby an over pressurization of the aquifer.

Mineral trapping, which is the slowest of the processes, is the final phase. It results from the geochemical reactions of carbonic acid (H_2CO_3) and the native minerals of the formation. This trapping mechanism is dependent on the rock minerals, the pressure of the gas, temperature and porosity of the host formation [85]. However, if mineral trapping is hastened, it may weaken the cap rock and the overlying formation causing a serious leak in CO_2. From above, it appears that the extent to which the CO_2 reacts with the formation water (dominated by its solubility) will vary according to factors such as pressure, temperature, the solubility of CO_2, the fluid and fluid/rock chemistry. The selection of a proper MO-NF could enhance the trapping mechanisms, and ultimately ensure an efficient CO_2 sequestration.

This is potentially achieved by injecting a nanofluid that buffers the acidity within the host formation (**Figure 14b**), but more importantly will yield a gel-like material (**Figure 14a**), denser than the resident fluid in the host formation.

In this study, it was found that formulating a nanofluid from Si-NP and polyvinyl alcohol under CO_2 bubbling would lead to the formation of silicated gels. Increasing the load in Si-NP yields a rigid gel (**Figure 15**).

The results suggest that condensation of SiO_2-NF depends rather on the load in Si-NP than the concentration in PVOH. However, further investigations are required to understand the extent to which the host formation-fluid chemistry alters the solubility of CO_2, and the host formation parameters (fluid chemistry, temperature, and pressure) alter the gel formation.

4.3 Renewable energy production

4.3.1 Overview of photo thermal energy production

Up to date, the primary energy supplied for human needs comes from fossil and nuclear resources. These can be harmful to environment because they cause global warning, ozone layer depletion, biosphere and geosphere destruction, and ecological devastation [86]. These drawbacks have geared the attention towards cleaner energy. Solar energy is one of the most promising amongst them not only because its exploitation can fulfill the entire world demand in energy, but more importantly it is one of the cleanest source or energy [87].

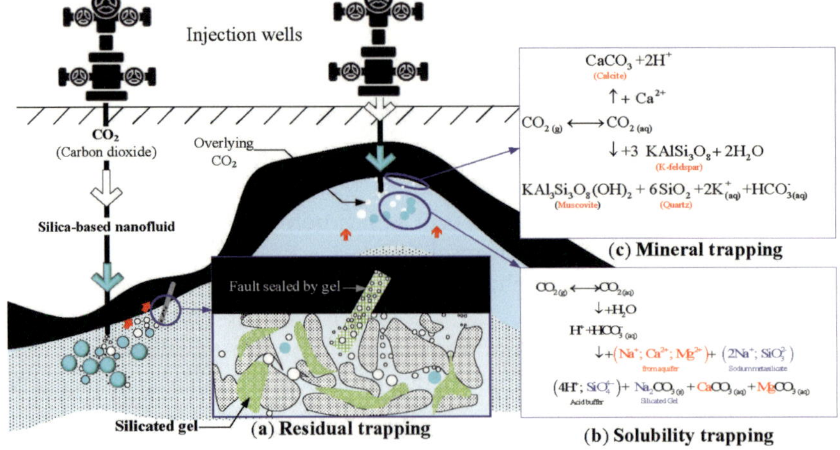

Figure 14.
Conceptual enhancement of CO_2 sequestration by Si-NF injection.

Figure 15.
Silicated-gel formed from Si-NP and polyvinyl alcohol without (a) porous medium and (b) in a porous medium.

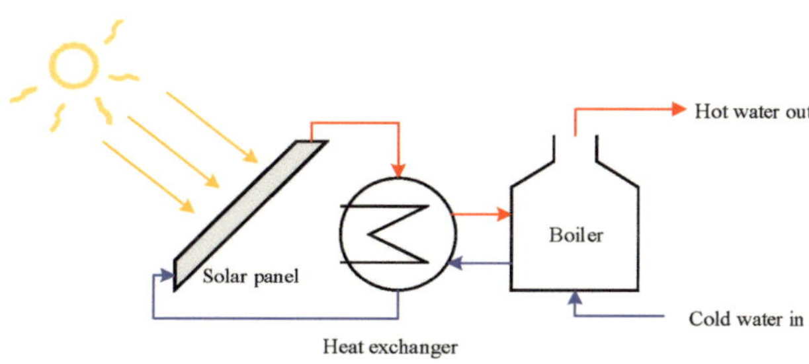

Figure 16.
Schematic representation of a thermal solar system for water heating applications.

Solar thermal energy consists in transforming photons into heat, which is collected and distributed by the means of devices called solar collectors [88]. A solar system usually has at least three components including the solar collector, a heat exchanger, and a thermal storage system (**Figure 16**).

Solar collectors use working fluids exposed to the solar irradiation. One of the most attractive candidates working fluid is water, given its price, its availability, and its eco-friendliness. The optimal temperature that can be obtained from a solar system depends on the type of solar collectors, which include flate-plate, parabolic through and parabolic dish collectors [89, 90].

4.3.2 Relevance of MO-NPs in solar thermal systems

The efficiency of a solar thermal systems is strongly related to the thermal efficiency of the working fluid, which is the ratio of useful energy effectively transferred. Ideal working fluids should exhibit peculiar thermophysical properties such as low viscosity, high thermal conductivity and high specific heat capacity, in addition to chemical stability [91]. The interest of MO-NPs for solar applications lies on their ability to enhance of the properties, which results in a significant improvement of the outlet temperature and the net power produced within a solar thermal system.

The literature reports that MO nanofluids have a higher thermal conductivity than that water taken alone [92]. The replacement of water as Heat Transfer Fluid (HTF) by a SiO_2-water based nanofluid has demonstrated the ability to enhance the performance of a FP collector through the reduction of the viscosity, the enhancement of the specific heat capacity and the rise of the temperature of the HTF at the outlet temperature of the solar system [93].

A theoretical model assessing the performance of a solar thermal system using Al_2O_3-water based nanofluid as Direct Absorber Solar Collector (DASC) has shown an increase of 10% of the collector efficiency compared to water-based flat plate solar collectors operating in the same conditions [94]. However, the optical performance of MO-NF as DASC depends of the volume fraction of MO-NPs [95]. MO-NP can also be associated with photovoltaic devices within hybrid Photovoltaic Thermal (PV/T) systems.

In such configurations, the electrical and thermal energy are simultaneously generated by a photovoltaic module and the working fluid, respectively [96]. Furthermore, nanofluids, exhibiting magnetic properties, offer the possibility to increase the thermal conductivity of a working fluid upon the application of a magnetic field [97, 98]. Nano ferrofluids, standing amongst, were found to improve significantly the efficiency of the photothermal or PV/T systems [99].

However, the major challenge with such types of nanofluids is the formulation of stable working fluid. This is so because nano ferrofluids have the propensity either to agglomerate in solution [100] or to suffer from chemical instability [101]. The combination of Fe_3O_4 with other MO-NP in composite materials usually allows to overcome those limitations. Therefore, exploration of the thermophysical features of composite MO nanofluids is an interesting direction aiming at the optimization of solar thermal systems.

5. Concluding remarks

The present chapter has covered the topic of metal oxide nanoparticles (MO-NP) and nanocomposites (MO-NCP) and their application for carbon capture utilization and storage (CCUS). It was shown that MO-NP (or MO-NCP) can be synthesized using physical, chemical, and biological methods. Each approach is tuned to give a MO-NP (or MO-NCP) with specific features. Regardless the production route, it was highlighted the stability of the nanofluid was the main challenge.

In respect of CCUS application, the most prominent results were obtained from silica (SiO_2) and alumina (Al_2O_3) oxides. SiO_2-NP could alter the wettability in a manner to increase the production of heavy oil. Dispersed into polymeric base fluid, it was shown SiO_2-NP could yield a gel-structure, which can plug the large voids of formation, leading to either an increment of oil production or prevent the leakage of sequestered gases. On the other, Al_2O_3-NP and its silicate composite could delay the deposition in heavy materials during the oil production using carbon dioxide (CO_2) injection. On the other hand, it was shown that ferrous and ferric nanofluids could improve the heat transfer of the fluid, making them good candidate for solar thermal energy.

Acknowledgements

The authors would extend their acknowledgement to Mitsubishi Chemical and Japan Petroleum Exploration for supplying respectively the polymers and the crude oils used in this study.

Conflict of interest

The authors declare no conflict of interest.

Funding

The works presented in this chapter were supported JSPS KAKENHI (Grant Number JP20K21163 and JP19K15490).

Author details

Ronald Nguele[1*], Katia Nchimi Nono[2] and Kyuro Sasaki[1]

1 Kyushu University, Fukuoka, Japan

2 University of Yaoundé I, Cameroon

*Address all correspondence to: nguele@mine.kyushu-u.ac.jp

References

[1] Coey JMD, Venkatesan M, Xu H. Introduction to Magnetic Oxides. Functional Metal Oxides. 2013. p. 1-49.

[2] Yuan C, Wu H Bin, Xie Y, Lou XW (David). Mixed Transition-Metal Oxides: Design, Synthesis, and Energy-Related Applications. Angew Chemie Int Ed [Internet]. 2014 Feb 3;53(6):1488-504.

[3] Fang J, Xuan Y, Li Q. Preparation of three-dimensionally ordered macroporous perovskite materials. Chinese Sci Bull. 2011 Jul 1;56:2156-2161.

[4] Sadakane M, Ueda W. Three-Dimensionally Ordered Macroporous (3DOM) Perovskite Mixed Metal Oxides. In: Perovskites and Related Mixed Oxides. Weinheim, Germany: Wiley-VCH Verlag GmbH & Co. KGaA; 2015. p. 113-142.

[5] Merkel TJ, Herlihy KP, Nunes J, Orgel RM, Rolland JP, DeSimone JM. Scalable, shape-specific, top-down fabrication methods for the synthesis of engineered colloidal particles. Langmuir. 2010 Aug;26(16):13086-13096.

[6] Gerberich WW, Jungk JM, Mook WM. The Bottom-Up Approach To Materials By Design. In: Meyers MA, Ritchie RO, Sarikaya MBT-N and MD of AM, editors. Oxford: Elsevier Science Ltd; 2003. p. 211-220.

[7] Zhao X, Zheng B, Li C, Gu H. Acetate-derived ZnO ultrafine particles synthesized by spray pyrolysis. Powder Technol. 1998;100(1):20-23.

[8] Azurdia J, Marchal J, Laine R. Synthesis and Characterization of Mixed-Metal Oxide Nanopowders Along the CoO_x–Al_2O_3 Tie Line Using Liquid-Feed Flame Spray Pyrolysis. J Am Ceram Soc. 2006 Sep 1;89:2749-2756.

[9] Azurdia JA, McCrum A, Laine RM. Systematic synthesis of mixed-metal oxides in NiO–Co3O4, NiO–MoO_3, and NiO–CuO systems via liquid-feed flame spray pyrolysis. J Mater Chem. 2008;18(27):3249-3258.

[10] Chao LT, Wei M, MacManus-Driscoll JL. Synthesis and characterisation of nanocrystalline iron oxides via ultrasonic spray assisted chemical vapour deposition. J Phys Conf Ser. 2006 Feb 22;26(1):304-307.

[11] Suslick KS. Sonochemistry. Science (80). 1990 Mar 23;247(4949):1439-45.

[12] Kumar RV, Diamant Y, Gedanken A. Sonochemical Synthesis and Characterization of Nanometer-Size Transition Metal Oxides from Metal Acetates. Chem Mater. 2000 Aug 1;12(8):2301-2305.

[13] Díez-García MI, Manzi-Orezzoli V, Jankulovska M, Anandan S, Bonete P, Gómez R, et al. Effects of Ultrasound Irradiation on the Synthesis of Metal Oxide Nanostructures. Phys Procedia. 2015;63:85-90.

[14] Shi W, Song S, Zhang H. Hydrothermal synthetic strategies of inorganic semiconducting nanostructures. Chem Soc Rev. 2013;42(13):5714-5743.

[15] Ota J, Srivastava SK. Polypyrrole Coating of Tartaric Acid-Assisted Synthesized Bi_2S_3 Nanorods. J Phys Chem C. 2007 Aug 1;111(33):12260-12264.

[16] Yuan G, Cao Y, Zan N, Schulz HM, Gluyas J, Hao F, et al. Coupled mineral alteration and oil degradation in thermal oil-water-feldspar systems and implications for organic-inorganic interactions in hydrocarbon reservoirs. Geochim Cosmochim Acta. 2019;248:61-87.

[17] Thapa R, Maiti S, Rana TH, Maiti UN, Chattopadhyay KK. Anatase TiO_2 nanoparticles synthesis via simple hydrothermal route: Degradation of Orange II, Methyl Orange and Rhodamine B. J Mol Catal A Chem. 2012;363-364:223-229.

[18] Xiaoming F. Synthesis and Optical Absorpition Properies of Anatase TiO_2 Nanoparticles via a Hydrothermal Hydrolysis Method. Rare Met Mater Eng. 2015 May;44(5):1067-1070.

[19] Tong H, Enomoto N, Inada M, Tanaka Y, Hojo J. Hydrothermal synthesis of mesoporous TiO_2-SiO_2 core-shell composites for dye-sensitized solar cells. Electrochim Acta. 2014 Jun 1;130:329-334.

[20] Lu J, Qi D, Deng C, Zhang X, Yang P. Hydrothermal synthesis of α-Fe_2O_3@SnO_2 core–shell nanotubes for highly selective enrichment of phosphopeptides for mass spectrometry analysis. Nanoscale . 2010;2(10):1892-1900.

[21] Parashar M, Shukla V, Singh R. Metal oxides nanoparticles via sol–gel method: a review on synthesis, characterization and applications. J Mater Sci Mater Electron. 2020 Mar;31.

[22] Li X-L, Peng Q, Yi J-X, Wang X, Li Y. Near monodisperse TiO_2 nanoparticles and nanorods. Chemistry. 2006;12(8):2383—2391.

[23] Zeng H, Rice PM, Wang SX, Sun S. Shape-Controlled Synthesis and Shape-Induced Texture of $MnFe_2O_4$ Nanoparticles. J Am Chem Soc. 2004 Sep 1;126(37):11458-11459.

[24] Chen JC, Chen W-C, Tien Y-C, Shih C-J. Effect of calcination temperature on the crystallite growth of cerium oxide nano-powders prepared by the co-precipitation process. J Alloy Compd. 2010 Apr;496:364-369.

[25] Fu C, Ravindra NM. Magnetic iron oxide nanoparticles: synthesis and applications. Bioinspired, Biomim Nanobiomaterials. 2012 Aug;1(4):229-244.

[26] Liu J, You D, Yu M, Li S. Preparation and characterization of hollow glass microspheres–cobalt ferrite core-shell particles based on homogeneous coprecipitation. Mater Lett. 2011;65(5):929-932.

[27] Ansari MA, Khan HM, Alzohairy MA, Jalal M, Ali SG, Pal R, et al. Green synthesis of Al_2O_3 nanoparticles and their bactericidal potential against clinical isolates of multi-drug resistant Pseudomonas aeruginosa. World J Microbiol Biotechnol. 2015 Jan 11;31(1):153-164.

[28] Senthilkumar S, Rajendran A. Biosynthesis of TiO_2 nanoparticles using Justicia gendarussa leaves for photocatalytic and toxicity studies. Res Chem Intermed. 2018;44(10):5923-5940.

[29] Fridman AA. Plasma chemistry. Cambridge; New York: Cambridge University Press; 2008.

[30] Petitpas G, Rollier J-D, Darmon A, Gonzalez-Aguilar J, Metkemeijer R, Fulcheri L. A comparative study of non-thermal plasma assisted reforming technologies. Int J Hydrogen Energy . 2007;32(14):2848-2867.

[31] Acayanka E, Tarkwa J-B, Nchimi KN, Voufouo SAY, Tiya-Djowe A, Kamgang GY, et al. Grafting of N-doped titania nanoparticles synthesized by the plasma-assisted method on textile surface for sunlight photocatalytic self-cleaning applications. Surfaces and Interfaces. 2019;

[32] Acayanka E, Kuete DS, Kamgang GY, Nzali S, Laminsi S, Ndifon PT. Synthesis, Characterization

and Photocatalytic Application of TiO2/SnO2 Nanocomposite Obtained Under Non-thermal Plasma Condition at Atmospheric Pressure. Plasma Chem Plasma Process. 2016 May 21;36(3):799-811.

[33] Acayanka E, Tarkwa J-B, Nchimi KN, Voufouo SAY, Tiya-Djowe A, Kamgang GY, et al. Grafting of N-doped titania nanoparticles synthesized by the plasma-assisted method on textile surface for sunlight photocatalytic self-cleaning applications. Surfaces and Interfaces. 2019 Dec;17:100361.

[34] Chemical Fundamentals of Hydrometallurgy. Hydrometallurgy. 2013. p. 21-64. (Wiley Online Books).

[35] Manivasakan P, Rajendran V, Rauta PR, Sahu BB, Panda BK. Direct Synthesis of Nano Alumina from Natural Bauxite. Adv Mater Res. 2009 Apr 1;67:143-148.

[36] Rayzman V, Aturin A, Pevzner I, Sizyakov V, Ni L, Filipovich I. Extracting Silica and Alumina from Low-Grade Bauxite. JOM. 2003 Jan 8;55:47-50.

[37] Akoh H, Tsukasaki Y, Yatsuya S, Tasaki A. Magnetic properties of ferromagnetic ultrafine particles prepared by vacuum evaporation on running oil substrate. J Cryst Growth. 1978 Dec 1;45:495-500.

[38] Phuoc TX, Soong Y, Chyu MK. Synthesis of Ag-deionized water nanofluids using multi-beam laser ablation in liquids. Opt Lasers Eng. 2007 Dec;45(12):1099-1106.

[39] Everett DH. Chapter 2. Why are Colloidal Dispersions Stable? I Basic Principles. In 1988. p. 16-29.

[40] Bolukbasi A, Ciloglu D. Pool boiling heat transfer characteristics of vertical cylinder quenched by SiO_2–water nanofluids. Int J Therm Sci. 2011 Jun 1;50(6):1013-1021.

[41] Darzi AAR, Farhadi M, Sedighi K, Shafaghat R, Zabihi K. Experimental investigation of turbulent heat transfer and flow characteristics of SiO_2/water nanofluid within helically corrugated tubes. Int Commun Heat Mass Transf. 2012 Nov 1;39(9):1425-1434.

[42] Kumar RS, Sharma T. Stability and rheological properties of nanofluids stabilized by SiO_2 nanoparticles and SiO_2-TiO_2 nanocomposites for oilfield applications. Colloids Surfaces A Physicochem Eng Asp. 2018 Feb;539:171-183.

[43] Nguele R, Sreu T, Inoue H, Sugai Y, Sasaki K. Enhancing Oil Production Using Silica-Based Nanofluids: Preparation, Stability, and Displacement Mechanisms. Ind Eng Chem Res. 2019 Aug;58(32):15045-15060.

[44] Suslick KS, Didenko Y, Fang MM, Hyeon T, Kolbeck KJ, McNamara III WB, et al. Acoustic cavitation and its consequences. Philos Trans R Soc A. 1999;357(1927):335-353.

[45] Xuan Y, Li Q. Heat transfer enhancement of nanofluids. Int J Heat Fluid Flow. 2000 Feb 1;21(1):58-64.

[46] Garg J, Poudel B, Chiesa M, Gordon JB, Ma JJ, Wang JB, et al. Enhanced thermal conductivity and viscosity of copper nanoparticles in ethylene glycol nanofluid. J Appl Phys. 2008 Apr 2;103(7):074301.

[47] Ngo I, Sasaki K, Nguele R, Sugai Y. Formation Damage Induced by Water-Based Alumina Nanofluids during Enhanced Oil Recovery: Influence of Postflush Salinity. ACS Omega. 2020 Oct 27;5(42):27103-27112.

[48] Ali N, Teixeira JA, Addali A. A Review on Nanofluids: Fabrication, Stability, and Thermophysical Properties. J Nanomater. 2018; 2018:1-33.

[49] Kato H, Nakamura A, Noda N. Determination of size distribution of silica nanoparticles: A comparison of scanning electron microscopy, dynamic light scattering, and flow field-flow fractionation with multiangle light scattering methods. Mater Express. 2014;4(2):144-152.

[50] Sidik NAC, Mohammed HA, Alawi OA, Samion S. A review on preparation methods and challenges of nanofluids. Int Commun Heat Mass Transf. 2014 May;54:115-125.

[51] Fazeli SA, Hosseini Hashemi SM, Zirakzadeh H, Ashjaee M. Experimental and numerical investigation of heat transfer in a miniature heat sink utilizing silica nanofluid. Superlattices Microstruct. 2012 Feb 1;51(2):247-264.

[52] Pang C, Jung J-Y, Lee JW, Kang YT. Thermal conductivity measurement of methanol-based nanofluids with Al_2O_3 and SiO_2 nanoparticles. Int J Heat Mass Transf. 2012 Oct 1;55(21-22):5597-5602.

[53] Devendiran DK, Amirtham VA. A review on preparation, characterization, properties and applications of nanofluids. Renew Sustain Energy Rev. 2016;60:21-40.

[54] Shah A, Fishwick R, Wood J, Leeke G, Rigby S, Greaves M. A review of novel techniques for heavy oil and bitumen extraction and upgrading. Energy Environ Sci. 2010;3(6):700.

[55] Or C, Sasaki K, Sugai Y, Nakano M, Imai M. Preliminary numerical modelling of CO_2 gas foaming in heavy oil and simulations of oil production from heavy oil reservoirs. Can J Chem Eng. 2016 Mar;94(3):576-585.

[56] Speight JG. Petroleum asphaltenes - Part 1: Asphaltenes, resins and the structure of petroleum. Oil Gas Sci Technol. 2004;59(5):467-477.

[57] Nguele R, Ghulami MR, Sasaki K, Said-Al Salim H, Widiatmojo A, Sugai Y, et al. Asphaltene Aggregation in Crude Oils during Supercritical Gas Injection. Energy and Fuels. 2016 Feb;30(2):1266-1278.

[58] Nguele R, Sasaki K, Ghulami MR, Sugai Y, Nakano M. Pseudo-phase equilibrium of light and heavy crude oils for enhanced oil recovery. J Pet Explor Prod Technol. 2016 Sep 20;6(3):419-432.

[59] Ngo I, Srisuriyachai F, Sasaki K, Sugai Y, Nguele R. Effects of Reversibility on Enhanced Oil Recovery Using Sodium Dodecylbenzene Sulfonate (SDBS). J Japan Pet Inst. 2019 Jul 1;62(4):188-198.

[60] Nguele R, Sasaki K, Salim HS, Sugai Y. Physicochemical and microemulsion properties of dimeric quaternary ammonium salts with trimethylene spacer for enhanced oil recovery. Colloid Polym Sci. 2015 Dec 2;293(12):3487-3497.

[61] Nguele R, Sasaki K, Sugai Y, Said Al-Salim H, Ueda R. Mobilization and displacement of heavy oil by cationic microemulsions in different sandstone formations. J Pet Sci Eng. 2017 Aug;157:1115-1129.

[62] Santanna VC, Curbelo FDS, Castro Dantas TN, Dantas Neto AA, Albuquerque HS, Garnica AIC. Microemulsion flooding for enhanced oil recovery. J Pet Sci Eng. 2009 Jun;66(3-4):117-120.

[63] Nguele R, Sasaki K, Salim HS-A, Sugai Y, Widiatmojo A, Nakano M. Microemulsion and phase behavior properties of (Dimeric ammonium surfactant salt – heavy crude oil – connate water) system. J Unconv Oil Gas Resour. 2016 Jun;14:62-71.

[64] Bera A, Mandal A. Microemulsions: a novel approach to enhanced oil

recovery: a review. J Pet Explor Prod Technol. 2015 Sep;5(3):255-268.

[65] Thomas S. Enhanced oil recovery-an overview. Oil Gas Sci Technol. 2008;63(1):9-19.

[66] Ansah EO, Sugai Y, Nguele R, Sasaki K. Integrated microbial enhanced oil recovery (MEOR) simulation: Main influencing parameters and uncertainty assessment. J Pet Sci Eng. 2018 Dec;171:784-793.

[67] Purwasena IA, Astuti DI, Syukron M, Amaniyah M, Sugai Y. Stability test of biosurfactant produced by Bacillus licheniformis DS1 using experimental design and its application for MEOR. J Pet Sci Eng. 2019 Dec 1;183.

[68] Ansah EO, Vo Thanh H, Sugai Y, Nguele R, Sasaki K. Microbe-induced fluid viscosity variation: field-scale simulation, sensitivity and geological uncertainty. J Pet Explor Prod Technol. 2020 Jun 1;10(5):1983-2003.

[69] Ogolo NA, Olafuyi OA, Onyekonwu MO. Enhanced Oil Recovery Using Nanoparticles. In: SPE Saudi Arabia Section Technical Symposium and Exhibition. Society of Petroleum Engineers; 2012.

[70] Hendraningrat L, Li S, Torsæter O. A coreflood investigation of nanofluid enhanced oil recovery. J Pet Sci Eng. 2013 Nov;111:128-138.

[71] Li S, Hendraningrat L, Torsaeter O. Improved Oil Recovery by Hydrophilic Silica Nanoparticles Suspension: 2-Phase Flow Experimental Studies. In: International Petroleum Technology Conference. International Petroleum Technology Conference; 2013.

[72] Giraldo J, Benjumea P, Lopera S, Cortés FB, Ruiz MA. Wettability Alteration of Sandstone Cores by Alumina-Based Nanofluids. Energy & Fuels. 2013 Jul;27(7):3659-3665.

[73] Kothari N, Raina B, Chandak KB, Iyer V, Mahajan HP. Application Of Ferrofluids For Enhanced Surfactant Flooding In IOR. In: SPE EUROPEC/EAGE Annual Conference and Exhibition. Society of Petroleum Engineers; 2010.

[74] Tarek M, El-Banbi AH. Comprehensive Investigation of Effects of Nano-Fluid Mixtures to Enhance Oil Recovery. In: SPE North Africa Technical Conference and Exhibition. Society of Petroleum Engineers; 2015.

[75] Haroun MR, Alhassan S, Ansari AA, Al Kindy NAM, Abou Sayed N, Abdul Kareem BA, et al. Smart Nano-EOR Process for Abu Dhabi Carbonate Reservoirs. In: Abu Dhabi International Petroleum Conference and Exhibition. Society of Petroleum Engineers; 2012. p. 1-13.

[76] Donaldson EC, Tiab DECD, Donaldson EC. Petrophysics: Theory and Practice of Measuring Reservoir Rock and Fluid Transport Properties. 2nd ed. Book. Gulf Professional Pub./Elsevier; 2004. 898 p.

[77] Tola S, Sasaki K, Sugai Y. Wettability alteration of sandstone with zinc oxide nano-particles. In: 23rd Formation Evaluation Symposium of Japan 2017. 2017.

[78] Romero Z, Disney R, Acuna HM, Cortes F, Patino JE, Cespedes Chavarro C, et al. Application and evaluation of a nanofluid containing nanoparticles for asphaltenes inhibition in well CPSXL4. In: OTC Brasil. Offshore Technology Conference; 2013.

[79] Ibrahim HH, Idem RO. Interrelationships between asphaltene precipitation inhibitor effectiveness, asphaltenes characteristics, and precipitation behavior during n-heptane (light paraffin hydrocarbon)-induced

asphaltene precipitation. Energy and Fuels. 2004 Jul;18(4):1038-1048.

[80] Goual L. Petroleum Asphaltenes. In: Manar El-Sayed Abdul-Raouf, editor. Crude Oil Emulsions- Composition Stability and Characterization. InTech; 2012. p. 27-42.

[81] Salmawati S, Sasaki K, Sugai Y, Yousefi-Sahzabi A. Estimating a baseline of soil CO_2 flux at CO_2 geological storage sites. Environ Monit Assess. 2019 Sep 14;191(9):563.

[82] Yousefi-Sahzabi A, Sasaki K, Yousefi H, Pirasteh S, Sugai Y. GIS aided prediction of CO_2 emission dispersion from geothermal electricity production. J Clean Prod. 2011 Nov 1;19(17-18):1982-1993.

[83] Sasaki K, Susanto V, Anggara F, Yousefi-Sahzabi A, Sugai Y, Kawamura T, Et Al. Few Considerations on Problems of CO_2 Geological Storage with Carbon Circulation and Proposal of An Integrated Regional Energy System considering Low Carbon Society. J MMIJ. 2015 Aug 1 [cited 2019 Sep 20];131(8_9):503-8.

[84] Ajayi T, Gomes JS, Bera A. A review of CO2 storage in geological formations emphasizing modeling, monitoring and capacity estimation approaches. Pet Sci [Internet]. 2019 Jul 8 [cited 2019 Sep 19];1-36. Available from: http://link.springer.com/10.1007/s12182-019-0340-8

[85] Ansah EO, Nguele R, Sugai Y, Sasaki K. Predicting the antagonistic effect between albite-anorthite synergy and anhydrite on chemical enhanced oil recovery: effect of inorganic ions and scaling. J Dispers Sci Technol. 2020 Dec 31;42(1):21-32.

[86] Serrano E, Rus G, García-Martínez J. Nanotechnology for sustainable energy. Vol. 13, Renewable and Sustainable Energy Reviews. Pergamon; 2009. p. 2373-2384.

[87] Kannan N, Vakeesan D. Solar energy for future world: - A review. Renew Sustain Energy Rev. 2016;62:1092-1105.

[88] Tian Y, Zhao CY. A review of solar collectors and thermal energy storage in solar thermal applications. Appl Energy. 2013;104:538-553.

[89] Reddy KS, Kamnapure NR, Srivastava S. Nanofluid and nanocomposite applications in solar energy conversion systems for performance enhancement: a review. Int J Low-Carbon Technol. 2017 Mar;12(1):1-23.

[90] Tchanche BF, Lambrinos G, Frangoudakis A, Papadakis G. Low-grade heat conversion into power using organic Rankine cycles – A review of various applications. Renew Sustain Energy Rev. 2011;15(8):3963-3979.

[91] Axaopoulos PJ. Solar Thermal Conversion: Active Solar Systems. Simmetria Publications; 2011.

[92] Lee S, Choi SU-S, Li S, Eastman JA. Measuring Thermal Conductivity of Fluids Containing Oxide Nanoparticles. J Heat Transfer. 1999 May 1;121(2):280-289.

[93] Akhatov JS, Mirzaev SZ, Halimov AS, Telyaev SS, Juraev ET. Study of the possibilities of thermal performance enhancement of flat plate solar water collectors by using of nanofluids as heat transfer fluid. Appl Sol Energy. 2017;53(3):250-257.

[94] Tyagi H, Phelan P, Prasher R. Predicted Efficiency of a Low-Temperature Nanofluid-Based Direct Absorption Solar Collector. J Sol Energy Eng Asme - J Sol Energy Eng. 2009 Nov;131.

[95] Karami M, Akhavan-Bahababadi MA, Delfani S, Raisee M. Experimental investigation of CuO nanofluid-based

Direct Absorption Solar Collector for residential applications. Renew Sustain Energy Rev. 2015;52:793-801.

[96] Michael JJ, Iniyan S. Performance analysis of a copper sheet laminated photovoltaic thermal collector using copper oxide – water nanofluid. Sol Energy. 2015;119:439-451.

[97] Shima PD, Philip J. Tuning of Thermal Conductivity and Rheology of Nanofluids Using an External Stimulus. J Phys Chem C. 2011 Oct 20;115(41):20097-20104.

[98] Lajvardi M, moghimi rad J, Hadi I, Gavili A, Isfahani T, Zabihi F, et al. Experimental investigation for enhanced ferrofluid heat transfer under magnetic field effect. J Magn Magn Mater. 2010 Nov;322:3508-13.

[99] Ghadiri M, Sardarabadi M, Pasandideh-fard M, Moghadam AJ. Experimental investigation of a PVT system performance using nano ferrofluids. Energy Convers Manag. 2015;103:468-476.

[100] Smith MJ, Ho VHB, Darton NJ, Slater NKH. Effect of Magnetite Nanoparticle Agglomerates on Ultrasound Induced Inertial Cavitation. Ultrasound Med Biol. 2009;35(6):1010-1014.

[101] Lagoeiro Le. Transformation of magnetite to hematite and its influence on the dissolution of iron oxide minerals. J Metamorph Geol. 1998 May;16(3):415-423.

Chapter 9

Effective Parameters on Increasing Efficiency of Microscale Heat Sinks and Application of Liquid Cooling in Real Life

Yousef Alihosseini, Amir Rezazad Bari and Mehdi Mohammadi

Abstract

Over the past two decades, electronic technology and miniaturization of electronic devices continue to grow exponentially, and heat dissipation becomes a critical issue for electronic devices due to larger heat generation. So, the need to cool down electronic components has led to the development of multiple cooling methods and microscale heat sinks. This chapter reviewed recent advances in developing an efficient heat sink, including (1) geometry parameters, (2) flow parameters that affect the hydraulic–thermal performance of the heat sink. Also, the main goal of this chapter is to address the current gap between academic research and industry. Furthermore, commercialized electronic cooling devices for various applications are highlighted, and their operating functions are discussed, which has not been presented before.

Keywords: microchannel, micro pin-fin, heat sink, heat transfer, microfluidic, commercialized microscale heat sink, electronic cooling

1. Introduction

Seminal advances in microelectronics technology have driven the Integrated Circuit Topographies (ICT) revolution over the last decades. Technologies of miniaturization, fabrication, and integrated circuit/system design are three vital parameters that have underpinned this revolution and allowed continuous and ongoing breakthroughs. However, the heat generated by electronic devices is always a fundamental problem that forced researchers to improve cooling systems to increase thermal efficiency. Since 85 °*c* is a critical temperature for electronic devices [1], exceed each 1 °*c* above critical temperature causes the reduction of 5% of devices life [2]. There are several methods to cool electronic devices as working fluid that are generally divided into (i) air cooling and (ii) liquid cooling. The efficiency of heat sinks increases due to the high thermal conductivity of liquids compared to air. Also, the increasing surface-to-volume ratio in heat sink leads to higher heat dissipation and extension of the electronic device's lifetime. Tuckerman and Pease [3] studied liquid cooling microchannel in single and multi-phase for the first time. Several parameters have also been considered to improve microchannel heat sinks efficiency, such as changing the cross-sections, patterns, manifolds, and working fluids [4].

Some technical issues have been reported, like generating hotspots and pressure drop through the microchannel for different applications. For instance, Copeland et al. [5] illustrated the impact of pressure drop and temperature gradient on system functionality. Moreover, they reported high-pressure drops (2 bars) for reaching minimal thermal resistance due to the small size of channels. Although utilizing a pump could compensate, the generated pressure drop which is used in conventional applications, using these pumps on a micro-scale is almost impossible [6]. The thermal boundary layer in convectional channels is maintained in a fully-developed state; thus, the thermal resistance increases and caused non-uniform heat transfer performance, leading to an unreliable platform and system failure. A large number of researches have been carried out to address these limitations by changing geometrical parameters and fluid flow structures in a microchannel.

In the current study, all previously reported parameters relating to enhancing the heat sink efficiency are considered. The efficient parameters on the performance of micro heat sinks are divided into two main parts, i.e., (i) Geometrical and (ii) Flow parameters. Geometrical parameters include patterns, cross-sections, and manifolds of heat sinks that the prior studies in this area are sorted and are explained in detail, and a comprehensive table is presented for each section. Also, working fluids (nano-fluids, phase change materials (PCMs) slurries, and boiling flows) are investigated as subsections of flow parameters. Besides, almost all micro heat sink applications in real life are characterized and the most significant of them, such as PCs and laptops, PCRs, gaming consoles, and data servers, are explained in detail, and other applications are listed. Finally, the suggestions and future direction of heat sink research are presented.

2. Microchannel and micro pin-fin heat sinks

The difference between macro-, mini-, and microchannels remains a lack of complete definition. However, it is fair to say that these differences can be classified into two groups; phenomenology and dimensions. Forces and phenomena play an essential role in micro-scale rather than macro-, mini scales [7]. Channel classification is based on hydraulic diameter as a simple guide to examine the desired dimensional range. Kandlikar and Grande [8] presented a general scheme based on the channel dimensions shown in **Table 1**.

In **Table 1**, D is the hydraulic diameter of the channel. In non-circular channels, it is recommended to use the smallest channel dimension in place of hydraulic diameter (e.g., the short side of a rectangular cross-section) [8]. Also, multiple microfluidic fabrication techniques have been developed, such as photolithography

Macrochannels	**$D > 3mm$**
Minichannels	**$3mm \geq D \geq 200\ \mu m$**
Microchannels	**$200\ \mu m \geq D \geq 10\ \mu m$**
Transitional microchannels	**$10\ \mu m \geq D \geq 1\ \mu m$**
Transitional nanochannels	**$1 \mu m \geq D \geq 0.1\ \mu m$**
Nanochannels	**$D \geq 0.1\ \mu m$**

Table 1.
Channel classification [8].

and soft lithography [9], laser cutting [10, 11], 3D printing [12, 13], microinjection molding [14] and glass etching for different applications like Point of Care (POC) and diagnosis [15, 16], microbiology [17, 18], drug delivery [19–21], oil and gas [22], micropump [23, 24], particle separation and enrichment [25–27], Organ on a chip [28–30], biosensor [31–34].

Microchannel and micro pin-fin are two types of heat sinks that are used in electronic cooling systems. The microchannel heat sink consists of extended parallel channels in different cross-sections (such as rectangular, hexagonal, triangular, etc.) that coolant flow passes from channels and absorbs heat from the chip. With advances in nano/micro-manufacturing techniques, another type of heat sink used in cooling circuits is a micro pin-fin heat sink. This heat sink type consists of pin-fin arrays in different shapes (like rectangular, hexagonal, elliptic, circular, etc.) and due to the high flow mixing rate, thermal performance increases compared to the microchannel heat sink.

3. Effective parameters on the efficiency of heat sinks

The thermal/hydraulic performance of the heat sink is affected by geometrical parameters (such as the shape of the cross-section, pattern, inlet/outlet arrangement) and flow parameters (such as working fluids and boiling flow) [35]. In this section, the effective parameters on the heat sink's thermal–hydraulic performance are presented.

3.1 Geometrical parameters

3.1.1 Patterns

Previous research indicates that changing pattern plays a fundamental role in enhancing the heat transfer rate [4]. The concept of periodic renewal of thermal boundary layers is a useful technique for enhancing heat transfer. Besides, secondary flows and fluid mixing are considered other factors for heat transfer enhancement that can be formed in pattern design.

Furthermore, research has shown that increasing heat transfer will reduce the pressure drop penalty [2, 3, 36]. Therefore, setting a balance between the heat transfer enhancement and the pressure drop penalty is required for discovering the optimum pattern design. Some relations, such as efficiency index (η) and Performance Evaluation Criteria (PEC), could help to identify these crucial parameters [37].

Several works studied the impact of pattern designs on heat transfer including, periodic (wavy, zigzag, etc.) [38–42], serpentine [43, 44], pin-fin [45, 46], and oblique [47–50] and most efficient pattern designs are summarized in **Table 2**.

The impact of the microchannel heat sink's pattern on thermal performance was investigated numerically by Lin et al. [51]. They reported that due to dean vortices formation in the channel's cross-section, the fluid mixing enhanced, and the thermal boundary layers' thickness decreased. Therefore, wavy heat sinks had better thermal performance compared to conventional straight heat sink due to higher Nusselt number and lower thermal resistance. After Lin et al. [51], another research group, Sui et al. [38] investigated the effect of wave amplitudes in wavy microchannel shown in **Figure 1**. Results illustrated that with increasing the amplitude to wavelength ratio (relative waviness), the thermal performance increased compared to the straight microchannel.

Author	Type of heat sink	Pattern	Size of heat sink	Re/Flow rate	Heat flux/ Power input
Y.Sui et al. [38]	MCHS	Wavy & Straight pattern	N.A.	100–800	$q'' = 1.5 \times 10^6$ And 1.5×10^6 W/m^2
Lin Lin et al. [51]	MCHS	Wavy	W = 10 mm L = 14 mm	300–800	$q'' = 100$ W/cm^2
M. Khoshvaght-Aliabadi et al. [41]	MCHS	Zigzag pattern with rectangular, triangular, and circular nook	W = 20 mm L = 100 mm	10–900	N.A.
Ahmad F. Al-Neama et al. [43]	MCHS	Serpentine (SPSM, DPSM, TPSM) & Straight rectangular pattern	W = 45 mm L = 41 mm	Flow Rate= 0.1, 0.2 to 1 L/min	Heat power = 100 W
Yogesh K. Prajapati [52]	MCHS	rectangular plate-fin	W = 3.7 mm L = 15 mm	100–800	q'' = 100 to 500 (kW/m^2)
Mushtaq Ismael Hasan [53]	PFHS	Square, triangular and circular	W = 6 mm L = 16 mm	100–900	N.A.
Zohreh Chamanroy and Morteza Khoshvaght-Aliabadi [54]	MCHS	Interrupted straight and wavy: SMHS, SMHS-SP, SMHS-WP 1, SMHS-WP 2, WMHS, WMHS-SP, WMHS-WP 1, and WMHS-WP 2	W = 1 mm L = 100 mm	100–1000	N.A.
Lei Chai and Liang Wang [55]	MCHS	Five different configurations of ribs: rectangular, backward triangular, diamond, forward triangular and ellipsoidal	W = 0.25 mm L = 10 mm	100–700	q'' = 106 (W/m^2)
Dawei Yang et al. [56]	PFHS	Triangle, square, pentagon, hexagon and circle geometries	W = 10 mm L = 10 mm	Re = 2122	N.A.
Zekeriya Parlak [42]	MCHS	Zigzag, straight and wavy pattern	N.A.	Inlet velocity = 0.5 to 5 m/s	q'' =2 ×106 (W/m^2)
Fatima Zohra Bakhti and Mohammad Si-Ameur [57]	PFHS	circular perforated pin fin	N.A.	100–400	Heat power = 300 W
N. Pahlevaninejad et al. [58]	MCHS	wavy pattern with rectangular obstacles	N.A.	5, 50, 150 and 300	q'' =50000 (W/m^2)

Author	Type of heat sink	Pattern	Size of heat sink	Re/Flow rate	Heat flux/ Power input
Mir Waqas Alam et al. [59]	PFHS	triangular shape micro-pin-fin	W = 13 mm L = 21 mm	500 to 10,000	q″ =2 (kW/m^2)

Note. MCHS: microchannel heat sink, PFHS: pin-fin heat sink, L: Length, W: Width, N.A: Not Applicable/Not Available, SPSM: Single path serpentine microchannel, DPSM: Double path serpentine microchannel, TPSM: Triple path serpentine microchannel, SMHS: Smooth straight mini-channel heat sink, SMHS-SP: SMHS with straight pin-fins, SMHS-WP: SMHS with wavy pin-fins, WMHS: Smooth wavy mini-channel heat sink, WMHS-SP: WMHS with straight pin-fins, WMHS-WP: WMHS with wavy pin-fins.

Table 2.
Summary of studies in different types of patterns.

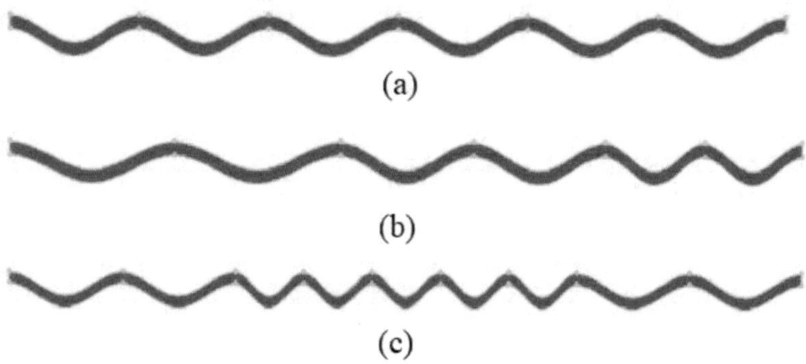

Figure 1.
Top view of wavy microchannels with (a) constant wavelength, (b) decreasing wavelength, and (c) shorter wavelength [38].

Mohammed et al. [60] numerically investigated the effect of different geometric patterns (zigzag, curvy, and step) on the heat transfer characteristics. The hydraulic- thermal characteristics (temperature profile, heat transfer coefficient, pressure drop, friction factor, and wall shear stress) were compared between considered patterns. They found that the highest heat transfer coefficient and the pressure drop belong to the zigzag, wavy, and curvy pattern, respectively. The step pattern obtains the lowest heat transfer coefficient and pressure drop; however, it was still higher than the conventional straight pattern. The main reason for the heat transfer enhancement is the periodic renewal of boundary layers. These boundary layers disturb by the formation of recirculation flow around the corners in the zigzag pattern and dean vortices' formation in the wavy and curvy patterns. Thus, the zigzag and the step were the best patterns for achieving the optimum hydraulic-thermal performance, respectively [42, 60].

Junye and Hugh Wang [61] numerically investigated the effect of different layout configurations on the flow distribution and pressure drop. They simulated six configurations, including single serpentine, multiple serpentines with two channels, multiple serpentines with three channels, multiple serpentines with six channels, straight parallel and interdigitated configurations with U-type arrangement for inlet/outlet position. They reported that less pressure drop and higher flow maldistribution (MLD) belong to a straight parallel configuration. Single serpentine had the best uniform flow distribution with a higher pressure drop, while multiple serpentine configurations had the medium pressure drop and flow MLD. Moreover, the flow MLD had decreased with the decreasing channel number. Similarly,

Al-Neama et al. [43] experimentally and numerically investigated the thermal performance of different serpentine patterns with straight rectangular microchannels (**Figure 2**). Results showed that the highest heat transfer belonged to Single path serpentine microchannel, followed by Double path serpentine microchannel and Triple path serpentine microchannel.

Many researchers in recent years studied the impact of different pin-fin patterns in heat sinks. A heat sink's performance can be enhanced by using different pin-fin patterns in which secondary flows and fluid mixing can be formed. The effects of using oblique fin pattern on the thermo-hydraulic performance of microchannel heat sink were studied by Yong-Jiun Le [45]. They reported that the oblique fin pattern gained a higher heat transfer performance enhancement of about 47% compared to the conventional microchannel due to the secondary flow generation and redevelopment of boundary layers.

Evaluation of thermal performance in hexagonal pin-fin heat sink was studied by S. Subramanian et al. [45]. The results revealed that the hexagonal fins achieved a higher heat transfer rate compared to the conventional straight microchannel. The significant reason for heat transfer enhancement is the formation of recirculation flow zone around the pin-fins, increasing the fluid mixing and disturbs the thermal boundary layer.

3.1.2 Cross-sections

The shape of the cross-section plays a vital role in heat sink performance. The micro pin-fin/channel cross-section can affect the flow characteristics like flow distribution, thermal resistance, secondary flow generation, maximum wall temperature, and thermal resistance, which can influence the heat transfer and pressure drop [8].

Generally, the cross-section shape is divided into two parts, (i) shape of microchannel cross-section (ii) shape of micro pin-fin cross-section; **Figure 3** shows different microchannel and micro pin-fins cross-sectional shape.

Gunnasegaran et al. [62] numerically investigated the effect of rectangular, triangular, and trapezoidal cross-section shapes on the microchannel heat transfer characteristics. The results showed that the rectangular cross-section gains the best maximum heat transfer coefficient of about 9.65 at the maximum Reynolds number (Re = 1000); while, the triangular shape showed the lowest heat transfer coefficient (9.38). In another study, Wang et al. [63] reported that the rectangular shape had the maximum, and the triangular shape had the minimum pressure drops.

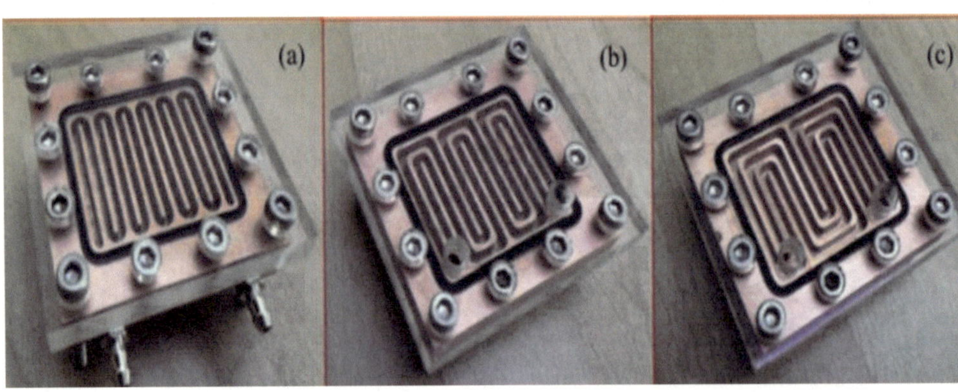

Figure 2.
Actual views of a serpentine microchannel with different layout configurations: (a) single, (b) double and (c) triple path multi-serpentine [43].

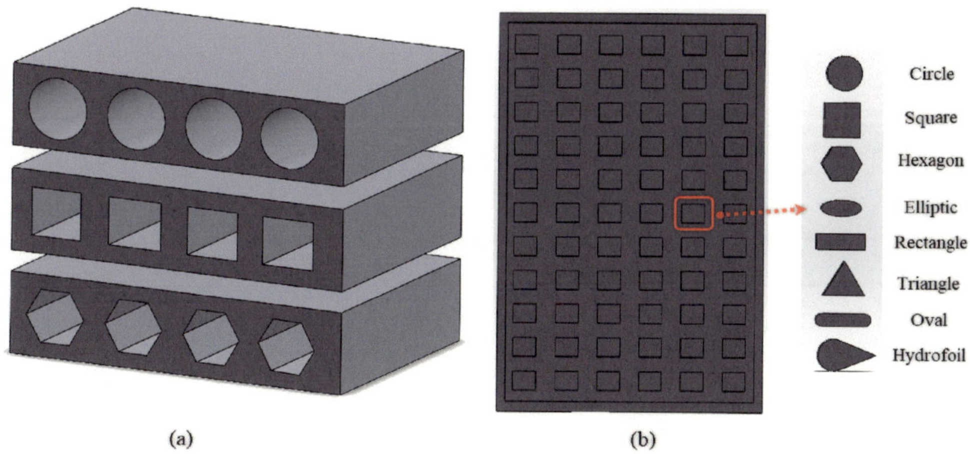

Figure 3.
(a) Shape of microchannel cross-section and (b) shape of micro pin-fin cross-section.

The effect of several aspect ratios on microchannel heat sinks' performance was studied numerically by Alfaryjat et al. [64]. They found that the highest heat transfer coefficient and the lowest pressure drop belong to the hexagonal and circular cross-section, respectively.

Evaluation of thermal performance in the micro-pin heat sink with various cross-section shapes was studied by Hasan [53]. The results revealed that the circular fins present a higher heat transfer rate compared to other fins. The square has the highest pressure drop, while it is the lowest for circular cross-section. Besides, in another research, Tehmina Ambreen et al. [65] numerically investigated the effect of different cross-section shapes (circular, square, and hexagonal) on the micro pin-fin heat sink thermal performance. Their results showed that the upstream fins row considerably influences the heat sink flow distribution and thermal performance. Moreover, circular fins showed the highest thermal performance (Nu_{ave} = 10), followed by hexagon and square fins, whereas square fins showed the smallest thermal performance values (Nu_{ave} = 7).

Dawei Yang et al. [56] illustrated that pin-fins with triangular cross-sections create the maximum blocking region in the fins back-side area, reducing the heat transfer rate of the heat sink increases the pressure drop. The circular cross-section had the minimum blocking region and pressure drop. Furthermore, the hexagonal shape has a flow-guided effect, which conducts the coolant into the back area; consequently, it has the best heat transfer performance.

Another evaluation of the pin-fin thermal–hydraulic performance of heat sinks belongs to Yang et al. [66], and their results showed that the sine shape's presented better heat transfer than hydrofoil and rhombus due to high fluid mixing. Besides, the rhombus has a maximum pressure drop because the flow path was smaller than other cases, while the pressure drop is minimum in the sine shape. Furthermore, due to the minimum stagnation area around the sine pin-fin heat sink, this one obtains the highest thermal performance. In a similar study, Ambreen et al. [67] investigated the effect of different micro-pin shapes on the heat sink's thermal performance. Results showed that the highest heat transfer rate belonged to circular pin-fin, followed by square and triangular shape. They demonstrated that the largest separation area happens behind the square and triangular pin-fins, and the circular had the least separation region, which contributes to the optimized thermal performance of the circular pin-fins (**Figure 4**). **Table 3** summarized the different types of cross-sections.

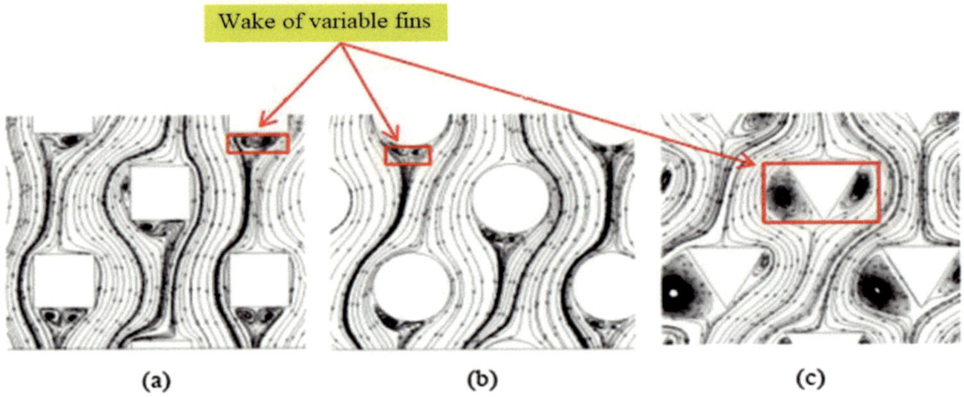

Figure 4.
Flow streamline of (a) square, (b) circular, and (c) triangular micro pin-fin [67].

Author	Type of heat sink	Size of a heat sink	Heat flux/ Power input	Type of shapes
Yanjun Zhang et al. [68]	Microchannel	W = 2 mm L = 20 mm	q″ = 35 × 105 (W/m^2)	Rectangular, Circular, Trapezoidal
A.A. Alfaryjat et al. [64]	Microchannel	W = 22 mm L = 12 mm	q″ = 500 (kW/m^2)	Hexagonal, Circular, Rhombus
Hamdi E. Ahmed and Mirghani I. Ahmed [69]	Microchannel	N.A.	q″ = 1000 (kW/m^2)	Trapezoidal, Triangular, Rectangular
Gongnan Xie et al. [70]	Microchannel	W = 35 mm L = 35 mm	Heat power = 300 W	Rectangular
Mushtaq Ismael Hasan [53]	Micro pin-fin	W = 6 mm L = 16 mm	N.A.	Square, Circular, Triangular
S. Subramanian et al. [45]	Micro pin-fin Plate pin-fin	W = 12.5 mm L = 25 mm	q″ = 32 × 104 (W/m^2)	Hexagonal fin Plate fin
Tehmina Ambreen and Man-Hoe Kim [65]	Micro pin-fin	W = 52.80 mm L = 125 mm	q″ = 37.2 (kW/m2)	Square, Circular, Hexagonal
Yong Jiun Lee et al. [49]	Microchannel	W = 12.7 mm L = 12.7 mm	Heat power = 273 W	Oblique fin
Dawei Yang et al. [56]	Micro pin-fin	W = 10.3 mm L = 10.3 mm	q″ = 144 (W/cm^2)	Rhombus, Hydrofoil, Sine
Fatima Zohra Bakhti and Mohamed Si-Ameur [57]	Micro pin-fin	N.A.	Heat power = 300 W	cylindrical perforated fins
Gagan V. Kewalramaniet al. [71]	Micro pin-fin	W = 10 mm L = 30.4 mm	q″ = 10, 50 and 100 (W/cm^2)	elliptical pin fin with different aspect ratio

Note. L: Length, W: Width, and N.A: Not Applicable/Not Available.

Table 3.
Summary of some studies in a different type of cross-sections.

According to results, a microchannel with a rectangular cross-section presented maximum performance compared to other microchannel cross-sections. The shape with no sharp corners obtains higher performance in micro pin-fins. It is hard to conclude precisely the best cross-section because the applied conditions play a significant role in micro pin-fin/channels heat sink performance.

3.1.3 Manifolds

Designing the manifolds is another primary geometrical parameter that researchers focus on to achieve a high-performance heat sink. Studying manifolds can be classified into three categories, including (a) location of inlet and outlet, (b) fluid inlet and outlet configuration (horizontal or vertical), and (c) header shape types. Some differences between experimental and theoretical results have been reported due to the maldistribution (MLD) in microchannel's branches and forming non-uniform temperature distribution in the edges of multiple microchannels. On the other hand, no difference was observed in single microchannels' results [72–75]. So, it can be concluded that an essential goal of studying manifolds is to achieve uniform flow and temperature distribution and remove hot spots for obtaining optimal performance.

Anbumeenakshi and Thansekhar [76] experimentally examined the effect of header shapes and inlet configurations in flow MLD in a rectangular microchannel heat sink (**Figure 5**). Results illustrated that trapezoidal and triangular types showed better flow uniformity at low flow rates. Also, the rectangular header improved flow MLD at high flow rates.

Xia et al. [77] analyzed the effects of three inlet and outlet flow arrangements (I, C, and Z-type), as well as header shapes (triangular, trapezoidal, and rectangular). The results illustrate that the I-type arrangement generated a uniform flow distribution compared to other configurations. Similarly, the rectangular header shape produced better flow uniformity than other headers. Critical parameters for flow distribution in the manifold are summarized in **Table 4**.

Note. H: Horizontal, V: Vertical, Min: Minimum, Max: Maximum, Rec: Rectangular, Trp: Trapezoidal, Tri: Triangular, MLD: Maldistribution, and N.A: Not Applicable/Not Available.

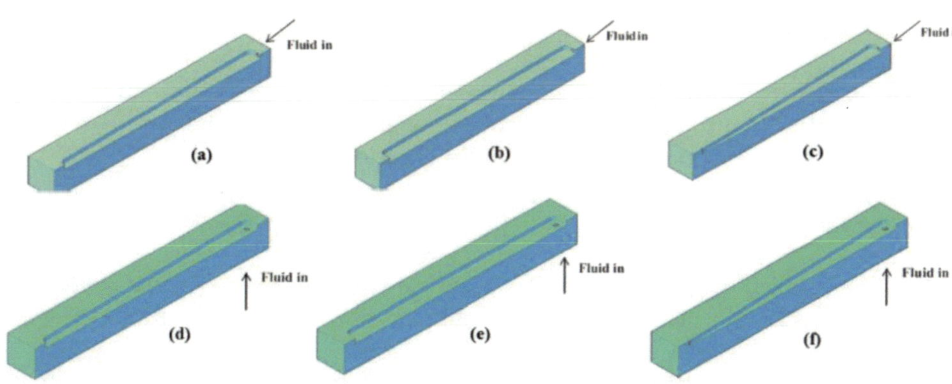

Figure 5.
Different header shape and inlet configurations. (a) Trapezoidal-inline. (b) Rectangular-inline. (c) Triangular-inline. (d) Trapezoidal-vertical. (e) Rectangular-vertical. (f) Triangular-vertical [76].

Author	Parameters	Inlet type	Outlet type	Max MLD	Min MLD
A.B. Datta et al. [78]	U/Z/Mixed-type	H/H/H	H/H/H	N.A.	Mixed
S.S. Sehgal et al. [79]	P/U/S-type	V/H/H	V/H/H	N.A.	S-type
Chi-Chuan wang et al. [80]	U/Z-type	H/H	H/H	Z-type	U-type

Author	Parameters	Inlet type	Outlet type	Max MLD	Min MLD
Junye Wang and Hualin Wang [81]	U/Z-type	H/H	H/H	Z-type	U-type
Guodong Xia et al. [77]	Tri./Trp./Rec. header shape & C/I/Z-type	V/V/V & V/V/V	V/V/V & V/V/V	Tri. & Z-type	Rec. & I-type
C.Anbumeenakshi et al. [76]	Rec./Trp./Tri. header shape	H/H/H & V/V/V	H/H/H & V/V/V	Rec. in low flow rate & Trp./Tri. In high flow rate	Rec. in high flow rate & Trp./Tri. in low flow rate
Chun-Kai Liu et al. [82]	Type of entrance flow	A: Side & B: front	A: Side & B: front	A	B
Kevin P. Drummond et al. [83]	hierarchical manifold	V	V	N.A.	N.A.
Wang Yabo et al. [84]	location of the inlet and outlet	V	V	N.A.	N.A.

Table 4.
Review literature about manifolds influence on flow distribution.

In order to reach the optimal flow distribution, maldistribution should be reduced along the microchannel. In most cases, Rectangular header shapes with vertical configurations cause low MLD compared to horizontal ones, and I-type has a symmetrical flow distribution that attracts most of the researchers for different applications.

3.2 Flow parameters

Flow parameter is another crucial parameter in the heat sinks thermal performance and according to the latest research, flow parameters can be classified into two categories: (1) working fluids and (2) Boiling flow.

3.2.1 Working fluids

The right selection of coolant fluid is a critical factor in removing the generated heat and according to literature, the air-cooling systems have received more attention in the past. However, with advances in electronic technology and the miniaturization of devices, conventional (air) cooling systems cannot remove more than 100 W/cm^2 generated heat by electronic devices. Therefore, liquid cooling such as water due to high specific heat capacity, high thermal conductivity, and more availability is considered more than ever [85]. Nowadays, with advances in nanotechnology, researchers classified working fluids into two categories, (1) nanofluids and (2) phase change material (PCM) slurries. They reported that improved thermophysical properties and higher thermal conductivity could be the advantages of these liquid coolants [86].

3.2.1.1 Nanofluids

Since fluids conductivity plays a vital role in the heat transfer from a solid surface to a fluid domain, conventional heat transfer coolant such as water, ethylene glycol, and paraffin have low thermal properties compared to metals and even metal oxides [87]. The use of nanoparticles is an effective method for modifying the heat transfer properties of fluids. Masuda et al. [88] were the first to study changes in the thermal conductivity and dynamic viscosity of base liquids with the additional fine metallic and non-metallic oxide particles.

The nanofluid and PCM's thermodynamical properties are defined based on the base fluid (i.e., DI water) [89].

Due to nanotechnology nanofluids advancement, different types of nanoparticles, nanotubes, and various distributed sizes were developed to investigate the stability and of nanofluidic during the cooling process [90].

3.2.1.2 Phase change material (PCM)

Advanced liquid coolants such as PCM are reported as effective substitutions for conventional coolants to enhance the heat transfer rate of microchannel heat sinks [91]. Furthermore, using phase change material (PCM) improves the coolants' thermophysical properties using the latent heat of melting.

PCM slurries are created by adding micro/nano encapsulated PCM particles to the base fluid (water, ethylene glycol, and paraffin). The PCM reveals a higher heat transfer rate when the PCM particles undergo a phase change transition [92–95]. One of the disadvantages of using nanofluids and PCM slurries is the higher viscosity than the base fluid, which imposes high pumping power on the system. Therefore, establishing a balance between heat transfer enhancement and pressure drop penalty is essential to distinguish the optimum advanced coolant [89]. Some of the significant nanofluids and PCM slurries used as working fluids are listed in **Table 5**.

Author	Phase one (base fluid)	Phase two (particles)
K.S. Suganthi et al. [96]	Ethylene glycol and water	ZnO as the nanoparticles
Zhou. Nianyong et al. [97]	Water	—
A.M. Bayomy et al. [98]	Water	—
Mingoo Choi and Keumnam Cho [99]	Water	5% Paraffin slurry as PCM
Bahram Rajabifar [86]	DI water	n-Octadecane as PCM and Alumina as nano particles
Jasim M. Mahdi and Emmanuel C. Nsoforet [100]	Water	RT82 as a PCM and $Al2O3$ as the nanoparticles
Lisi Jia et al. [101]	Water	TiO2 as the nanoparticles
Thaklaew Yiamsawas et al. [102]	Ethylene glycol and water	TiO2 and $Al2O3$ as the nanoparticles
Min Li [103]	Paraffin	Nano-graphite
Hamideh Sardarabadi et al. [104]	Water	Multi-walled carbon nanotubes
O. Pourmehran et al. [95]	Water	CuO as the nanoparticles
Arash Karimipour et al. [105]	Water	$Al2O3$ and AgO as the nanoparticles
Vivek Kumar and Jahar Sarkar [106]	Water	$Al2O3$–MWCNT

Table 5.
Different types of working fluids used in previous investigations.

3.2.2 Boiling flow

From the cooling performance perspective, two-phase flow boiling in microchannel heat sinks is more efficient than its single-phase equivalent. A temperature lower than the critical temperature and the coolant's boiling points provide sufficient factors for increasing heat transfer rate. Due to the nature of boiling and turbulent flow in the microchannel, while requiring a low rate of coolant flow and maintaining the wall temperatures relatively uniform, the boiling heat transfer coefficient is much higher than conventional systems [107]. The main problems in this type of system are higher pressure drop and instability of the system. Due to the biphasic flow inside the channels and the creation of larger bubbles, the higher pressure drop in the system may occur because of the channels' poor design.

Numerous studies have been conducted to explore the convection heat transfer characteristics of two-phase flow boiling in micro pin-fin/channel heat sinks in recent years. Wei Wan et al. [108] experimentally examined the effect of cross-section shape on flow boiling characteristics of micro pin-fin heat sinks. Four types of staggered micro pin-fins with different cross-section shapes, i.e., square, circular, diamond, and streamline, were tested in this study. Results showed that the square shape presents the higher boiling heat transfer, followed by circular and streamlined ones. The diamond micro pin-fins presented the smallest pressure drop, while their main problem is instability at moderate to high heat fluxes. The streamline micro pin-fins presented the largest two-phase pressure drop. Besides, the square and circular micro pin-fins showed their superiority in reducing two-phase flow instabilities. Matthew Law and Poh-Seng Lee [49] conducted an experimental study of flow boiling heat transfer and pressure characteristics in straight-finned and oblique-finned microchannels. They reported that the oblique-finned microchannels' thermal performance is higher than straight-finned ones due to the increase in the density of bubbles generated in the convective boiling regime. The high pressure-drop in oblique-finned microchannels causes a sudden change in the flow orientation, where the fluid is being forced to flow through secondary channels. The pressure drop fluctuations in the oblique-finned pattern are much lower compared to the straight-finned; consequently, the pressure instabilities in the oblique-finned microchannels are relatively smaller than the straight-finned microchannels.

4. Applications of heat sinks in real life

Today, we can observe all levels of technological devices in everyday life, such as smartphones, gaming consoles, PCs, and laptops, etc. Companies changed many parameters to achieve the best performance of devices, including decreasing thickness, enhancing GPU and CPU numbers, and changing the appearance [109].

Based on the last researches, with the increasing number of microprocessors, the heat generation is intensified; thus, heat dissipation becomes a critical issue for thermal researchers. The manufacturer's primary purpose is to achieve the best performance of devices by minimizing device and heat sink size and increasing the heat transfer rate utilizing the different items such as changing the geometrical parameters and working fluids.

In this section, some commercial electronic cooling systems and their operating principles are discussed. The purpose of this section is to establish a link between the scientific and commercial platforms. All effective parameters discussed in previous sections are required to fabricate optimum microchannel/micro pin-fin heat sink.

4.1 Computers and laptops

High-performance computing is the ability to process data and perform complex calculations at high speeds. One of the most common questions any manufacturer receives from customers is, "Why does my laptop or computer generate so much heat?" Heat is a normal byproduct of computer operation; for instance, a high-performance computer generates considerable heat than a lower performance computer. So, if the computer cannot disburse its heat, it may overheat, and device life will reduce. Therefore, this is a significant challenge for the manufacturer to solve. A heat sink is a thermally conductive device placed over a CPU or GPU to absorb some generated heat. Faster and multi-core processors require more high-efficiency heat sinks to keep their temperatures within acceptable levels. Nowadays, some companies professionally work in cooling system designing fields.

Asus [110], Alphacool [111], CoolIT Systems [112], Cool Innovations Inc. [113], Xtreme Performance Gear (XPG) [114], Coolermaster [115], Antec [116], Swiftech [117] and Thermaltake [118] are popular companies that work in cooling field. According to their operating principles, all companies used water as a cooling fluid. **Figure 6** shows some of the micro pin-fin/channel heat sinks manufactured by cooling companies.

As shown in **Figure 6a–c**, some companies use a straight pattern microchannel, and others use a pin-fin pattern (presented in **Figure 6d** and **e**). Although complex patterns increase the heat transfer rate, they cause a significant pressure drop. Some companies use different patterns without considering the pressure drop penalty, such as a micro pin-fin pattern to enhance heat sink performance.

According to Section 3.1.3, fluid and temperature MLD are decreased by using I-type flow arrangements and vertical inlet/outlet type. It concludes that the vertical inlet/outlet and I-type arrangements are implemented in commercialized products. Some of the heat sinks are made by several manufacturers and use vertical inlet/outlet type. Asus ROG GX700 is the first liquid-cooled gaming laptop in the world and the liquid coolant is circulated around the heat-generating components like GPU by pumps. The coolant then heads back to the cooling module, where two heat sinks (radiators) help dissipate heat, that is shown in **Figure 7a**.

4.2 Gaming consoles

Many consoles were designed with cooling fans in the past, but due to advances in thermal engineering, novel consoles are designed with liquid cooling fins and heat sinks to keep the internal components safe from high generated heat. PlayStation 5 (PS5) is the new video game console developed by Sony [119] Interactive Entertainment that utilized a heat sink with a hard copper plate with

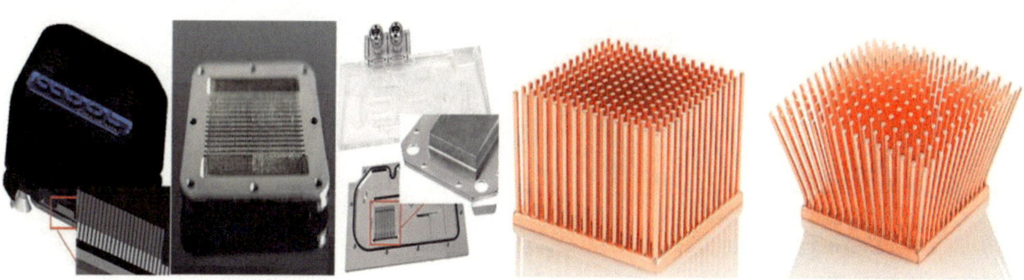

(a) (b) (c) (d) (e)

Figure 6.
(a) Alphacool, (b) Swiftech, (c) Thermaltake, (d) and (e) cool innovations Inc.

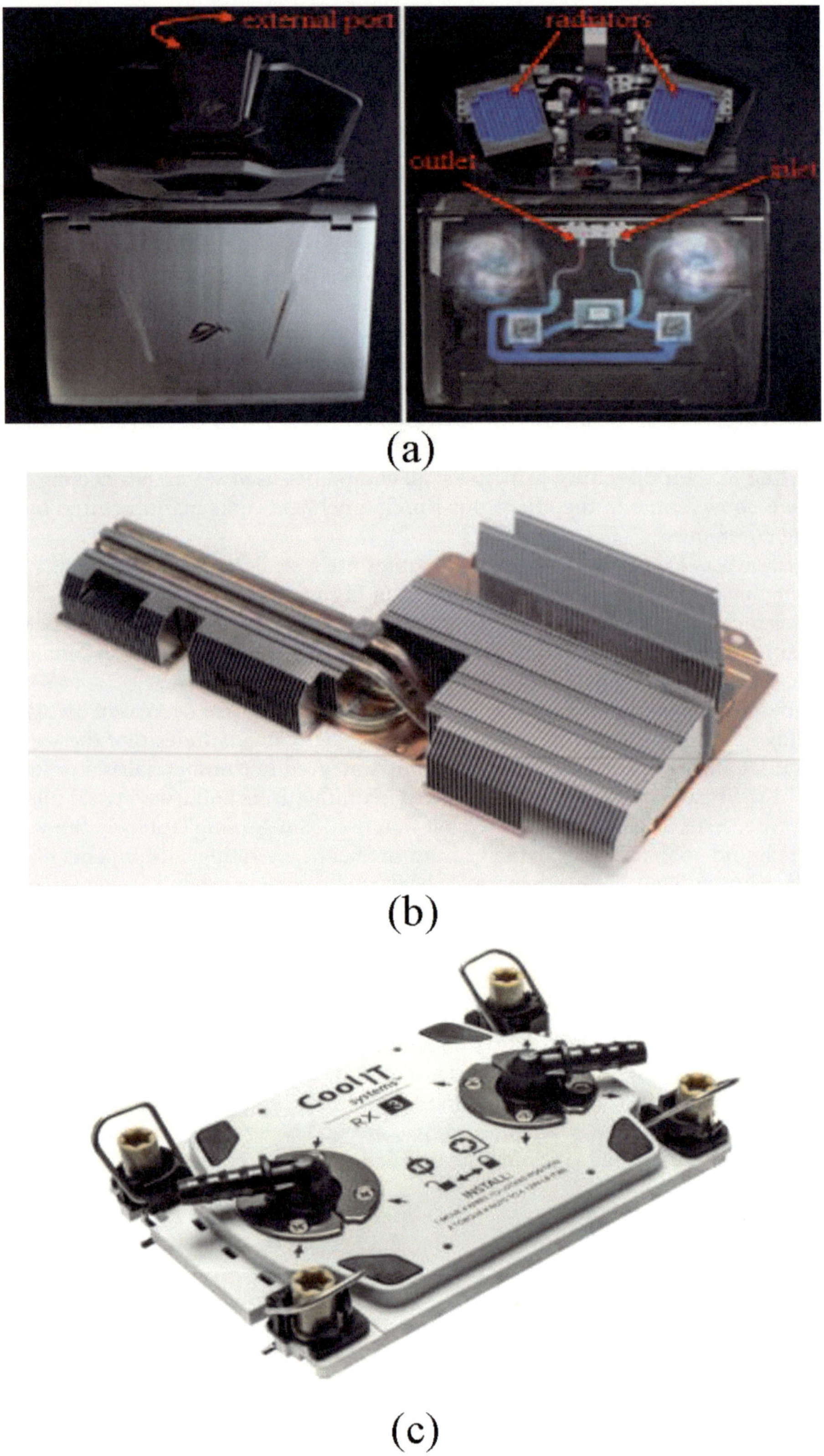

Figure 7.
(a) Asus ROG GX700 cooling system [110], (b)PlayStation 5 (PS5) heat sink type [119], (c) CPU RX3 [112].

aluminum dissipation fins, shown in **Figure 7b**. The generated heat from the chips was conducted into the heat sink, which used a heat pipe to move the heat across the fins.

4.3 Data server

High-performance servers are specially designed to handle large computational loads with fast data processing however, to reach higher speeds, the power dissipation of high-performance CPUs was challenging [112].

Due to the increase in generated heat by high-performance servers, air cooling systems could not dissipate this generated heat because of their lower thermal efficiency. Therefore, new methods of cooling (i.e., liquid cooling) should be tested. Coolit system [112] is one of the companies working in the data server cooling field that utilized Coolant Distribution Unit (CDU) instead of high failure rate components such as internal pumps to circulate coolant, see in **Figure 7c**.

4.4 Other applications

Wherever cooling is required, liquid cooling capabilities can be used to improve the system's thermal performance instead of traditional cooling methods. Liquid cooling systems are widely used to dissipate the heat generated by process operations in many applications, such as Aerospace systems, All-in-One devices, Food Industry instruments, and Biology fields.

In aerospace fields, thermal engineers have a major concern for keeping the plane at a steady temperature while avoiding ice buildup problems and protecting sensitive electronic components from extreme heat and cold. For that reason, companies should manufacture a range of heat sinks to keep planes and electronics components at optimal performance.

As shown in **Figure 7a**, an external port is required to cool down the All-in-one devices such as laptops and tablets, which may cause problems in the transportation and thermal performance of these devices. New liquid cooling methods can be used in these devices to optimize performance and facilitate transportation. It uses a small micro pump to recirculate a coolant in the integrated closed-circuit heat sink to dissipate heat away. Finally, heated liquid flows through the peltiers (Thermoelectric modules) to cool down and return to the circuit.

In traditional Food Industry fields, the lineup of food processing chillers works with air-cooled systems. Air-cooled food processing chillers use ambient air to dissipate heat from food cooling processes, which have low efficiency. New liquid cooling systems chillers use water from an external cooling tower to remove heat from food processes. These systems are long-lasting, quiet, and feature energy-efficiency properties. These cooling devices are best suited for medium temperature food processes, such as cheese, meat, and sauce production; potable liquid cooling systems tend to be economical solutions that deliver excellent cooling efficiency.

In biological fields, cooling devices are a common tool used in research labs. A thermal cycler (also known as a PCR device) is a laboratory instrument that facilitates DNA amplification through the polymerase chain reaction. PCR typically requires 20 to 35 cycles comprising two to three temperature steps. In the past, air-coolers used to control the heating and cooling process; by developing in the thermal engineering field, liquid cooling methods can be involved in thermal cycling to optimize the thermal efficiency and increase the number of tests per unit time [120].

5. Future research directions

Finally, according to the reviewed papers and commercialized devices, some research gaps recognized and the following directions are suggested:

I. Combination of micro pin-fin and oblique patterns to obtain novel patterns that may increase thermo-hydraulic performances

II. Investigation of hybrid nanofluids and PCMs in commercialized heat sinks can be an exciting field

III. Study of flow containing phase change particles and hybrid nanofluid within microchannels

IV. Using the hybrid pattern (micro pin-fin + microchannel) in commercialized heat sinks for increased thermal performance.

V. Using jet impingement heat sink that can remove high heat fluxes

VI. Using liquid cooling in small digital devices such as tablets and cellphones

6. Conclusion

Liquid cooling techniques in heat sinks have been using to increase the heat transfer rate in electronic devices. The performance of the heat sink is affected by some parameters. Geometry parameters (shape of the cross-section, pattern, manifolds) and flow parameters (working fluids and boiling) are the most common methods for enhancing the thermal–hydraulic performance of micro pin-fin/microchannel heat sink. Effective parameters are divided into two parts Geometry parameters and Flow parameters.

I. Flow parameters:

a. pattern: It was found that changing the channel flow path leads to generating the phenomena such as secondary flows in the channel cross-section that result in improving thermal performance. The previous data demonstrated, the geometric parameters play a significant role in heat sinks' performance

b. Cross-section: rectangular cross-section presents maximum performance compared to other microchannels heat sinks. While in micro pin-fins, it is impossible to conclude precisely the best cross-section due to the different operating conditions.

c. Manifold: it concludes that flow and temperature maldistribution should be reduced along the microchannel to reach the ideal flow distribution for obtaining optimal performance. Also, other parameters are crucial as well, such as header shape, inlet configurations (vertical and horizontal flow inlet/outlet), inlet/outlet arrangements (I, Z, C, S, V, and U-types), location of inlet/outlet.

II. Flow parameters:

a. Working fluid: since fluids conductivity plays a role in the convective heat transfer, conventional heat transfer coolant such as water, ethylene glycol, and paraffin have low thermal properties compared to metals and even metal oxides. Many studies reported that the heat transfer capability could be further enhanced by using nanofluids (ZnO, $Al2O3$, TiO2, etc.) and PCM (Paraffin slurry, n-Octadecane, etc.) due to their better heat transfer characteristics.

b. Boiling: flow boiling heat transfer in microchannels has been a subject of wide interest due to its ability to dissipate high heat fluxes; however, using this method still has critical challenges.

Author details

Yousef Alihosseini[1], Amir Rezazad Bari[1] and Mehdi Mohammadi[2,3]*

1 Faculty of Mechanical Engineering, Tarbiat Modares University, Tehran, Iran

2 Department of Mechanical and Manufacturing Engineering, University of Calgary, Calgary, Alberta, Canada

3 Biological Science, University of Calgary, Calgary, Alberta, Canada

*Address all correspondence to: Mehdi.mohammadiashan@ucalgary.ca

References

[1] Zhang, H.Y., Y.C. Mui, and M. Tarin, *Analysis of thermoelectric cooler performance for high power electronic packages*. Applied thermal engineering, 2010. **30**(6-7): p. 561-568.

[2] Ahmed, H.E., et al., *Optimization of thermal design of heat sinks: A review*. International Journal of Heat and Mass Transfer, 2018. **118**: p. 129-153.

[3] Tuckerman, D.B. and R.F.W. Pease, *High-performance heat sinking for VLSI*. IEEE Electron device letters, 1981. **2**(5): p. 126-129.

[4] Alihosseini, Y., et al., *Effect of a micro heat sink geometric design on thermo-hydraulic performance: A review*. Applied Thermal Engineering, 2020. **170**: p. 114974.

[5] Copeland, D., H. Takahira, and W. Nakayama, *MAN: FOLD M! CROCHANNEL HEAT S: NKS: THEORY AND EXPER: MENT*. 1995.

[6] Fan, Y., et al., *Thermal transport in oblique finned micro/minichannels*. 2015: Springer.

[7] Kew, P.A. and K. Cornwell, *Correlations for the prediction of boiling heat transfer in small-diameter channels*. Applied thermal engineering, 1997. **17** (8-10): p. 705-715.

[8] Kandlikar, S., et al., *Heat transfer and fluid flow in minichannels and microchannels*. 2005: elsevier.

[9] Madadi, H., et al., *A novel fabrication technique to minimize poly (dimethylsiloxane)-microchannels deformation under high-pressure operation*. Electrophoresis, 2013. **34** (22-23): p. 3126-32.

[10] Kinahan, D.J., et al., *Automation of Silica Bead-based Nucleic Acid Extraction on a Centrifugal Lab-on-a-Disc Platform*. Journal of Physics: Conference Series, 2016. **757**.

[11] Mohammadi, M., D.J. Kinahan, and J. Ducrée, *Lumped-Element Modeling for Rapid Design and Simulation of Digital Centrifugal Microfluidic Systems*, in *Lab-on-a-Chip Fabrication and Application*. 2016.

[12] Jang, T.-S., et al., *3D printing of hydrogel composite systems: Recent advances in technology for tissue engineering*. International Journal of Bioprinting, 2018. **4**(1).

[13] Manzanares Palenzuela, C.L. and M. Pumera, *(Bio)Analytical chemistry enabled by 3D printing: Sensors and biosensors*. TrAC Trends in Analytical Chemistry, 2018. **103**: p. 110-118.

[14] Attia, U.M., S. Marson, and J.R. Alcock, *Micro-injection moulding of polymer microfluidic devices*. Microfluidics and Nanofluidics, 2009. **7** (1): p. 1-28.

[15] Mohammadi, M., H. Madadi, and J. Casals-Terre, *Microfluidic point-of-care blood panel based on a novel technique: Reversible electroosmotic flow*. Biomicrofluidics, 2015. **9**(5): p. 054106.

[16] Mohammadi, M., et al., *Hydrodynamic and direct-current insulator-based dielectrophoresis (H-DC-iDEP) microfluidic blood plasma separation*. Anal Bioanal Chem, 2015. **407**(16): p. 4733-44.

[17] Khater, A., et al., *Dynamics of temperature-actuated droplets within microfluidics*. Sci Rep, 2019. **9**(1): p. 3832.

[18] Khater, A., et al., *Picoliter agar droplet breakup in microfluidics meets microbiology application: numerical and experimental approaches*. Lab on a Chip, 2020. **20**(12): p. 2175-2187.

[19] Saadat, M., et al., *Magnetic particle targeting for diagnosis and therapy of lung cancers.* Journal of Controlled Release, 2020. **328**: p. 776-791.

[20] Azizian, P., et al., *Electrohydrodynamic formation of single and double emulsions for low interfacial tension multiphase systems within microfluidics.* Chemical Engineering Science, 2019. **195**: p. 201-207.

[21] Agiotis, L., et al., *Magnetic manipulation of superparamagnetic nanoparticles in a microfluidic system for drug delivery applications.* Journal of Magnetism and Magnetic Materials, 2016. **401**: p. 956-964.

[22] Zhang, Y., et al., *Functionalized multiscale visual models to unravel flow and transport physics in porous structures.* Water Res, 2020. **175**: p. 115676.

[23] Dehghan Manshadi, M.K., et al., *Electroosmotic micropump for lab-on-a-chip biomedical applications.* International Journal of Numerical Modelling: Electronic Networks, Devices and Fields, 2016. **29**(5): p. 845-858.

[24] Madadi, H., J. Casals-Terre, and M. Mohammadi, *Self-driven filter-based blood plasma separator microfluidic chip for point-of-care testing.* Biofabrication, 2015. 7(2): p. 025007.

[25] Khetani, S., M. Mohammadi, and A. S. Nezhad, *Filter-based isolation, enrichment, and characterization of circulating tumor cells.* Biotechnol Bioeng, 2018. **115**(10): p. 2504-2529.

[26] Dehghan Manshadi, M.K., et al., *Induced-charge electrokinetics (ICEK) in microfluidics: A Review on Recent Advancements.* Journal of Micromechanics and Microengineering, 2020.

[27] Manshadi, M.K.D., et al., *Manipulation of micro- and nanoparticles in viscoelastic fluid flows within microfluid systems.* Biotechnol Bioeng, 2020. **117**(2): p. 580-592.

[28] Yong, K.W., et al., *Mesenchymal Stem Cell Therapy for Ischemic Tissues.* Stem Cells International, 2018. **2018**: p. 1-11.

[29] Soleimani, S., et al., *Translational models of tumor angiogenesis: A nexus of in silico and in vitro models.* Biotechnol Adv, 2018.

[30] Verhulsel, M., et al., *A review of microfabrication and hydrogel engineering for micro-organs on chips.* Biomaterials, 2014. **35**(6): p. 1816-32.

[31] Sevda Mohammadi, R.N., Mehdi Mohammadi Ashani, Hamid Sadabadi, Amir Sanati-Nezhad, Mohammad H. Zarifi, *Real-time monitoring of Escherichia coli concentration with planar microwave resonator sensor.* Microw Opt Technol Lett., 2019: p. 1–6.

[32] Salahandish, R., et al., *Reproducible and Scalable Generation of Multilayer Nanocomposite Constructs for Ultrasensitive Nanobiosensing.* Advanced Materials Technologies, 2019. **4**(11): p. 1900478.

[33] Narang, R., et al., *Sensitive, Real-time and Non-Intrusive Detection of Concentration and Growth of Pathogenic Bacteria using Microfluidic-Microwave Ring Resonator Biosensor.* Scientific Reports, 2018. **8**(1).

[34] Xu, L., et al., *Optical, electrochemical and electrical (nano)biosensors for detection of exosomes: A comprehensive overview.* Biosensors and Bioelectronics, 2020. **161**: p. 112222.

[35] Hussien, A.A., M.Z. Abdullah, and A.-N. Moh'd A, *Single-phase heat transfer enhancement in micro/minichannels using nanofluids: theory and applications.* Applied energy, 2016. **164**: p. 733-755.

[36] Kandlikar, S.G., *History, advances, and challenges in liquid flow and flow boiling heat transfer in microchannels: a critical review.* Journal of heat transfer, 2012. **134**(3).

[37] Rakhsha, M., et al., *Experimental and numerical investigations of turbulent forced convection flow of nano-fluid in helical coiled tubes at constant surface temperature.* Powder Technology, 2015. **283**: p. 178-189.

[38] Sui, Y., et al., *Fluid flow and heat transfer in wavy microchannels.* International Journal of Heat and Mass Transfer, 2010. **53**(13-14): p. 2760-2772.

[39] Mohammed, H.A., P. Gunnasegaran, and N.H. Shuaib, *Numerical simulation of heat transfer enhancement in wavy microchannel heat sink.* International Communications in Heat and Mass Transfer, 2011. **38**(1): p. 63-68.

[40] Sui, Y., P.S. Lee, and C.J. Teo, *An experimental study of flow friction and heat transfer in wavy microchannels with rectangular cross section.* International journal of thermal sciences, 2011. **50** (12): p. 2473-2482.

[41] Khoshvaght-Aliabadi, M., et al., *Effects of nooks configuration on hydrothermal performance of zigzag channels for nanofluid-cooled microelectronic heat sink.* Microelectronics Reliability, 2017. **79**: p. 153-165.

[42] Parlak, Z., *Optimal design of wavy microchannel and comparison of heat transfer characteristics with zigzag and straight geometries.* Heat and Mass Transfer, 2018. **54**(11): p. 3317-3328.

[43] Al-Neama, A.F., et al., *An experimental and numerical investigation of the use of liquid flow in serpentine microchannels for microelectronics cooling.* Applied Thermal Engineering, 2017. **116**: p. 709-723.

[44] Al-Neama, A.F., et al., *An experimental and numerical investigation of chevron fin structures in serpentine minichannel heat sinks.* International Journal of Heat and Mass Transfer, 2018. **120**: p. 1213-1228.

[45] Subramanian, S., K.S. Sridhar, and C.K. Umesh, *Experimental investigation of microchannel heat sink with modified hexagonal fins.* Journal of Applied Fluid Mechanics, 2019. **12**(3): p. 647-655.

[46] Duangthongsuk, W. and S. Wongwises, *A comparison of the heat transfer performance and pressure drop of nanofluid-cooled heat sinks with different miniature pin fin configurations.* Experimental Thermal and Fluid Science, 2015. **69**: p. 111-118.

[47] Lee, Y.J., P.K. Singh, and P.S. Lee, *Fluid flow and heat transfer investigations on enhanced microchannel heat sink using oblique fins with parametric study.* International Journal of Heat and Mass Transfer, 2015. **81**: p. 325-336.

[48] Law, M., P.-S. Lee, and K. Balasubramanian, *Experimental investigation of flow boiling heat transfer in novel oblique-finned microchannels.* International Journal of Heat and Mass Transfer, 2014. **76**: p. 419-431.

[49] Law, M. and P.-S. Lee, *A comparative study of experimental flow boiling heat transfer and pressure characteristics in straight-and oblique-finned microchannels.* International Journal of Heat and Mass Transfer, 2015. **85**: p. 797-810.

[50] Alihosseini, Y., M.Z. Targhi, and M. M. Heyhat, *Thermo-hydraulic performance of wavy microchannel heat sink with oblique grooved finned.* Applied Thermal Engineering, 2021: p. 116719.

[51] Lin, L., et al., *Heat transfer enhancement in microchannel heat sink by wavy channel with changing wavelength/*

amplitude. International Journal of Thermal Sciences, 2017. **118**: p. 423-434.

[52] Prajapati, Y.K., *Influence of fin height on heat transfer and fluid flow characteristics of rectangular microchannel heat sink.* International Journal of Heat and Mass Transfer, 2019. **137**: p. 1041-1052.

[53] Hasan, M.I., *Investigation of flow and heat transfer characteristics in micro pin fin heat sink with nanofluid.* Applied thermal engineering, 2014. **63**(2): p. 598-607.

[54] Chamanroy, Z. and M. Khoshvaght-Aliabadi, *Analysis of straight and wavy miniature heat sinks equipped with straight and wavy pin-fins.* International Journal of Thermal Sciences, 2019. **146**: p. 106071.

[55] Chai, L. and L. Wang, *Thermal-hydraulic performance of interrupted microchannel heat sinks with different rib geometries in transverse microchambers.* International Journal of Thermal Sciences, 2018. **127**: p. 201-212.

[56] Yang, D., et al., *Numerical and experimental analysis of cooling performance of single-phase array microchannel heat sinks with different pin-fin configurations.* Applied Thermal Engineering, 2017. **112**: p. 1547-1556.

[57] Bakhti, F.Z. and M. Si-Ameur, *A comparison of mixed convective heat transfer performance of nanofluids cooled heat sink with circular perforated pin fin.* Applied Thermal Engineering, 2019. **159**: p. 113819.

[58] Pahlevaninejad, N., M. Rahimi, and M. Gorzin, *Thermal and hydrodynamic analysis of non-Newtonian nanofluid in wavy microchannel.* Journal of Thermal Analysis and Calorimetry, 2020: p. 1-15.

[59] Alam, M.W., et al., *CPU heat sink cooling by triangular shape micro-pin-fin: Numerical study.* International Communications in Heat and Mass Transfer, 2020. **112**: p. 104455.

[60] Mohammed, H.A., P. Gunnasegaran, and N.H. Shuaib, *Influence of channel shape on the thermal and hydraulic performance of microchannel heat sink.* International Communications in Heat and Mass Transfer, 2011. **38**(4): p. 474-480.

[61] Wang, J. and H. Wang, *Flow-Field Designs of Bipolar Plates in PEM Fuel Cells: Theory and Applications.* Fuel Cells, 2012. **12**(6): p. 989-1003.

[62] Gunnasegaran, P., et al., *The effect of geometrical parameters on heat transfer characteristics of microchannels heat sink with different shapes.* International Communications in Heat and Mass Transfer, 2010. **37**(8): p. 1078-1086.

[63] Wang, H., Z. Chen, and J. Gao, *Influence of geometric parameters on flow and heat transfer performance of micro-channel heat sinks.* Applied Thermal Engineering, 2016. **107**: p. 870-879.

[64] Alfaryjat, A.A., et al., *Influence of geometrical parameters of hexagonal, circular, and rhombus microchannel heat sinks on the thermohydraulic characteristics.* International Communications in Heat and Mass Transfer, 2014. **52**: p. 121-131.

[65] Ambreen, T. and M.-H. Kim, *Effect of fin shape on the thermal performance of nanofluid-cooled micro pin-fin heat sinks.* International Journal of Heat and Mass Transfer, 2018. **126**: p. 245-256.

[66] Yang, D., et al., *Heat removal capacity of laminar coolant flow in a micro channel heat sink with different pin fins.* International Journal of Heat and Mass Transfer, 2017. **113**: p. 366-372.

[67] Ambreen, T., A. Saleem, and C.W. Park, *Pin-fin shape-dependent heat transfer and fluid flow characteristics of water-and nanofluid-cooled micropin-fin*

heat sinks: Square, circular and triangular fin cross-sections. Applied Thermal Engineering, 2019. **158**: p. 113781.

[68] Zhang, Y., S. Wang, and P. Ding, *Effects of channel shape on the cooling performance of hybrid micro-channel and slot-jet module.* International journal of heat and mass transfer, 2017. **113**: p. 295-309.

[69] Ahmed, H.E. and M.I. Ahmed, *Optimum thermal design of triangular, trapezoidal and rectangular grooved microchannel heat sinks.* International Communications in Heat and Mass Transfer, 2015. **66**: p. 47-57.

[70] Xie, G., et al., *Analysis of micro-channel heat sinks with rectangular-shaped flow obstructions.* Numerical Heat Transfer, Part A: Applications, 2016. **69** (4): p. 335-351.

[71] Kewalramani, G.V., et al., *Study of laminar single phase frictional factor and Nusselt number in In-line micro pin-fin heat sink for electronic cooling applications.* International Journal of Heat and Mass Transfer, 2019. **138**: p. 796-808.

[72] Choi, S.B., *Fluid flow and heat transfer in microtubes.* Micromechanical Sensors, Actuators, and Systems, ASME, 1991: p. 123-134.

[73] Peng, X.F., G.P. Peterson, and B.X. Wang, *Frictional flow characteristics of water flowing through rectangular microchannels.* Experimental Heat Transfer An International Journal, 1994. 7(4): p. 249-264.

[74] Harley, J.C., et al., *Gas Flow in Micr-Channels.* 1994.

[75] Cuta, J.M., C.E. McDonald, and A. Shekarriz, *Forced convection heat transfer in parallel channel array microchannel heat exchanger.* 1996, American Society of Mechanical Engineers, New York, NY (United States).

[76] Anbumeenakshi, C. and M.R. Thansekhar, *Experimental investigation of header shape and inlet configuration on flow maldistribution in microchannel.* Experimental Thermal and Fluid Science, 2016. **75**: p. 156-161.

[77] Xia, G., et al., *Effects of structural parameters on fluid flow and heat transfer in a microchannel with aligned fan-shaped reentrant cavities.* International Journal of Thermal Sciences, 2011. **50**(3): p. 411-419.

[78] Datta, A.B. and A.K. Majumdar, *A calculation procedure for two phase flow distribution in manifolds with and without heat transfer.* International Journal of Heat and Mass Transfer, 1983. **26**(9): p. 1321-1328.

[79] Sehgal, S.S., K. Murugesan, and S.K. Mohapatra, *Experimental investigation of the effect of flow arrangements on the performance of a micro-channel heat sink.* Experimental heat transfer, 2011. **24**(3): p. 215-233.

[80] Wang, C.-C., et al., *Characteristics of flow distribution in compact parallel flow heat exchangers, part I: typical inlet header.* Applied Thermal Engineering, 2011. **31**(16): p. 3226-3234.

[81] Wang, J. and H. Wang, *Discrete method for design of flow distribution in manifolds.* Applied Thermal Engineering, 2015. **89**: p. 927-945.

[82] Liu, C.-K., et al., *Effect of non-uniform heating on the performance of the microchannel heat sinks.* International communications in heat and mass transfer, 2013. **43**: p. 57-62.

[83] Drummond, K.P., et al., *A hierarchical manifold microchannel heat sink array for high-heat-flux two-phase cooling of electronics.* International Journal of Heat and Mass Transfer, 2018. **117**: p. 319-330.

[84] Wang, Y., et al., *Effects of the location of the inlet and outlet on heat*

transfer performance in pin fin CPU heat sink. Applied Thermal Engineering, 2019. **151**: p. 506-513.

[85] Saini, M. and R.L. Webb. *Heat rejection limits of air cooled plane fin heat sinks for computer cooling*. IEEE.

[86] Rajabifar, B., *Enhancement of the performance of a double layered microchannel heatsink using PCM slurry and nanofluid coolants.* International Journal of Heat and Mass Transfer, 2015. **88**: p. 627-635.

[87] Yu, W., et al., *Review and comparison of nanofluid thermal conductivity and heat transfer enhancements.* Heat transfer engineering, 2008. **29**(5): p. 432-460.

[88] Masuda, H., A. Ebata, and K. Teramae, *Alteration of thermal conductivity and viscosity of liquid by dispersing ultra-fine particles. Dispersion of Al2O3, SiO2 and TiO2 ultra-fine particles.* 1993.

[89] Joseph, M. and V. Sajith, *An investigation on heat transfer performance of polystyrene encapsulated n-octadecane based nanofluid in square channel.* Applied Thermal Engineering, 2019. **147**: p. 756-769.

[90] Maruf, S.H., et al., *Use of nanoimprinted surface patterns to mitigate colloidal deposition on ultrafiltration membranes.* Journal of membrane science, 2013. **428**: p. 598-607.

[91] Roberts, N.S., et al., *Efficacy of using slurry of metal-coated microencapsulated PCM for cooling in a micro-channel heat exchanger.* Applied Thermal Engineering, 2017. **122**: p. 11-18.

[92] Noh, N.M., A. Fazeli, and N.A.C. Sidik, *Numerical simulation of nanofluids for cooling efficiency in microchannel heat sink.* J. Adv. Res. Fluid Mech. Therm. Sci, 2014. **4**(1): p. 13-23.

[93] Wang, T., et al., *Thermal investigation of lithium-ion battery module with different cell arrangement structures and forced air-cooling strategies.* Applied energy, 2014. **134**: p. 229-238.

[94] Sharma, C.S., et al., *Energy efficient hotspot-targeted embedded liquid cooling of electronics.* Applied Energy, 2015. **138**: p. 414-422.

[95] Pourmehran, O., et al., *Numerical optimization of microchannel heat sink (MCHS) performance cooled by KKL based nanofluids in saturated porous medium.* Journal of the Taiwan Institute of Chemical Engineers, 2015. **55**: p. 49-68.

[96] Suganthi, K.S., V.L. Vinodhan, and K.S. Rajan, *Heat transfer performance and transport properties of ZnO–ethylene glycol and ZnO–ethylene glycol–water nanofluid coolants.* Applied energy, 2014. **135**: p. 548-559.

[97] Zhou, N., et al., *Experimental investigation on the performance of a water spray cooling system.* Applied Thermal Engineering, 2017. **112**: p. 1117-1128.

[98] Bayomy, A.M., M.Z. Saghir, and T. Yousefi, *Electronic cooling using water flow in aluminum metal foam heat sink: Experimental and numerical approach.* International Journal of Thermal Sciences, 2016. **109**: p. 182-200.

[99] Choi, M. and K. Cho, *Effect of the aspect ratio of rectangular channels on the heat transfer and hydrodynamics of paraffin slurry flow.* International journal of heat and mass transfer, 2001. **44**(1): p. 55-61.

[100] Mahdi, J.M. and E.C. Nsofor, *Solidification enhancement of PCM in a triplex-tube thermal energy storage system with nanoparticles and fins.* Applied Energy, 2018. **211**: p. 975-986.

[101] Jia, L., et al., *Improving the supercooling degree of titanium dioxide*

nanofluids with sodium dodecylsulfate. Applied energy, 2014. **124**: p. 248-255.

[102] Yiamsawas, T., et al., *Experimental studies on the viscosity of TiO2 and Al2O3 nanoparticles suspended in a mixture of ethylene glycol and water for high temperature applications.* Applied energy, 2013. **111**: p. 40-45.

[103] Li, M., *A nano-graphite/paraffin phase change material with high thermal conductivity.* Applied energy, 2013. **106**: p. 25-30.

[104] Sardarabadi, H., et al., *Experimental investigation of a novel type of two-phase closed thermosyphon filled with functionalized carbon nanotubes/ water nanofluids for electronic cooling application.* Energy Conversion and Management, 2019. **188**: p. 321-332.

[105] Karimipour, A., A. D'Orazio, and M.S. Shadloo, *The effects of different nano particles of Al2O3 and Ag on the MHD nano fluid flow and heat transfer in a microchannel including slip velocity and temperature jump.* Physica E: Low-Dimensional Systems and Nanostructures, 2017. **86**: p. 146-153.

[106] Kumar, V. and J. Sarkar, *Particle ratio optimization of Al2O3-MWCNT hybrid nanofluid in minichannel heat sink for best hydrothermal performance.* Applied Thermal Engineering, 2020. **165**: p. 114546.

[107] Mohammadi, A. and A. Koşar, *Review on heat and fluid flow in micro pin fin heat sinks under single-phase and two-phase flow conditions.* Nanoscale and Microscale Thermophysical Engineering, 2018. **22**(3): p. 153-197.

[108] Wan, W., et al., *Experimental study and optimization of pin fin shapes in flow boiling of micro pin fin heat sinks.* Applied Thermal Engineering, 2017. **114**: p. 436-449.

[109] Tang, H., et al., *Review of applications and developments of ultra-thin micro heat pipes for electronic cooling.* Applied energy, 2018. **223**: p. 383-400.

[110] Asus. *Products of Asus*. 2020; Available from: https://www.asus.com/.

[111] Alphacool. *Products of Alphacool*. 2020; Available from: https://www.alphacool.com/.

[112] CoolITsystem. *Products of CoolITsystem*. 2020; Available from: https://coolitsystems.com/.

[113] Coolinnovations. *Products of Coolinnovations*. 2020; Available from: http://www.coolinnovations.com/.

[114] XPG. *Products of XPG*. 2020; Available from: http://www.xpg.com/en/feature/553.

[115] Coolermaster. *Products of Coolermaster*. 2020; Available from: http://www.coolermaster.com/product/Lines/cpu-liquid-cooler/.

[116] Antec. *Products of Antec*. 2020; Available from: http://www.coolermaster.com/product/Lines/cpu-liquid-cooler/.

[117] Swiftech. *Products of Swiftech*. 2020; Available from: http://site.swiftech.com/.

[118] Thernaltake. *Products of Thermaltake*. 2020; Available from: https://www.thermaltake.com/cooler.aspx.

[119] Sony. *Sony Platstation*. 2020; Available from: https://www.playstation.com/.

[120] Khater, A., et al., *Thermal droplet microfluidics: from biology to cooling technology.* TrAC Trends in Analytical Chemistry, 2021: p. 116234.